Essential
Understanding
Series

Developing
Essential Understanding
of
Algebraic Thinking
for Teaching Mathematics in
Grades 3–5

D1517852

Maria Blanton
University of Massachusetts Dartmouth
Dartmouth, Massachusetts

Linda Levi
Teachers Development Group
West Linn, Oregon

Barbara J. Dougherty
Volume Editor
Iowa State University
Ames, Iowa

Terry Crites
Northern Arizona University
Flagstaff, Arizona

Rose Mary Zbiek
Series Editor
The Pennsylvania State University
University Park, Pennsylvania

Barbara J. Dougherty
Iowa State University
Ames, Iowa

NATIONAL COUNCIL OF
TEACHERS OF MATHEMATICS

Copyright © 2011 by
The National Council of Teachers of Mathematics, Inc.
1906 Association Drive, Reston, VA 20191-1502
(703) 620-9840; (800) 235-7566; www.nctm.org
All rights reserved

Third Printing 2016

Library of Congress Cataloging-in-Publication Data

Developing essential understanding of algebraic thinking for teaching
mathematics in grades 3-5 / Maria Blanton ... [et al.].
    p. cm. -- (Essential understanding series)
  Includes bibliographical references.
  ISBN 978-0-87353-668-4
  1. Algebra--Study and teaching (Elementary) 2. Problem solving. 3.
Critical thinking. I. Blanton, Maria L.
  QA159.D48 2011
  372.7--dc22
                              2011007755

The National Council of Teachers of Mathematics is a public voice of mathematics education,
supporting teachers to ensure equitable mathematics learning of the highest quality for all
students through vision, leadership, professional development, and research.

Printed in the United States of America

Contents

Foreword

Teaching mathematics in prekindergarten–grade 12 requires a special understanding of mathematics. Effective teachers of mathematics think about and beyond the content that they teach, seeking explanations and making connections to other topics, both inside and outside mathematics. Students meet curriculum and achievement expectations when they work with teachers who know what mathematics is important for each topic that they teach.

The National Council of Teachers of Mathematics (NCTM) presents the Essential Understanding Series in tandem with a call to focus the school mathematics curriculum in the spirit of *Curriculum Focal Points for Prekindergarten through Grade 8 Mathematics: A Quest for Coherence*, published in 2006, and *Focus in High School Mathematics: Reasoning and Sense Making*, released in 2009. The Essential Understanding books are a resource for individual teachers and groups of colleagues interested in engaging in mathematical thinking to enrich and extend their own knowledge of particular mathematics topics in ways that benefit their work with students. The topic of each book is an area of mathematics that is difficult for students to learn, challenging to teach, and critical for students' success as learners and in their future lives and careers.

Drawing on their experiences as teachers, researchers, and mathematicians, the authors have identified the big ideas that are at the heart of each book's topic. A set of essential understandings–mathematical points that capture the essence of the topic–fleshes out each big idea. Taken collectively, the big ideas and essential understandings give a view of a mathematics that is focused, connected, and useful to teachers. Links to topics that students encounter earlier and later in school mathematics and to instruction and assessment practices illustrate the relevance and importance of a teacher's essential understanding of mathematics.

On behalf of the Board of Directors, I offer sincere thanks and appreciation to everyone who has helped to make this series possible. I extend special thanks to Rose Mary Zbiek for her leadership as series editor. I join the Essential Understanding project team in welcoming you to these books and in wishing you many years of continued enjoyment of learning and teaching mathematics.

Henry Kepner
President, 2008–2010
National Council of Teachers of Mathematics

Preface

From prekindergarten through grade 12, the school mathematics curriculum includes important topics that are pivotal in students' development. Students who understand these ideas cross smoothly into new mathematical terrain and continue moving forward with assurance.

However, many of these topics have traditionally been challenging to teach as well as learn, and they often prove to be barriers rather than gateways to students' progress. Students who fail to get a solid grounding in them frequently lose momentum and struggle in subsequent work in mathematics and related disciplines.

The Essential Understanding Series identifies such topics at all levels. Teachers who engage students in these topics play critical roles in students' mathematical achievement. Each volume in the series invites teachers who aim to be not just proficient but outstanding in the classroom—teachers like you—to enrich their understanding of one or more of these topics to ensure students' continued development in mathematics.

How much do you need to know?

To teach these challenging topics effectively, you must draw on a mathematical understanding that is both broad and deep. The challenge is to know considerably more about the topic than you expect your students to know and learn.

Why does your knowledge need to be so extensive? Why must it go above and beyond what you need to teach and your students need to learn? The answer to this question has many parts.

To plan successful learning experiences, you need to understand different models and representations and, in some cases, emerging technologies as you evaluate curriculum materials and create lessons. As you choose and implement learning tasks, you need to know what to emphasize and why those ideas are mathematically important.

While engaging your students in lessons, you must anticipate their perplexities, help them avoid known pitfalls, and recognize and dispel misconceptions. You need to capitalize on unexpected classroom opportunities to make connections among mathematical ideas. If assessment shows that students have not understood the material adequately, you need to know how to address weaknesses that you have identified in their understanding. Your understanding must be sufficiently versatile to allow you to represent the mathematics in different ways to students who don't understand it the first time.

In addition, you need to know where the topic fits in the full span of the mathematics curriculum. You must understand where your students are coming from in their thinking and where they are heading mathematically in the months and years to come.

Accomplishing these tasks in mathematically sound ways is a tall order. A rich understanding of the mathematics supports the varied work of teaching as you guide your students and keep their learning on track.

How can the Essential Understanding Series help?

The Essential Understanding books offer you an opportunity to delve into the mathematics that you teach and reinforce your content knowledge. They do not include materials for you to use directly with your students, nor do they discuss classroom management, teaching styles, or assessment techniques. Instead, these books focus squarely on issues of mathematical content—the ideas and understanding that you must bring to your preparation, in-class instruction, one-on-one interactions with students, and assessment.

How do the authors approach the topics?

For each topic, the authors identify "big ideas" and "essential understandings." The big ideas are mathematical statements of overarching concepts that are central to a mathematical topic and link numerous smaller mathematical ideas into coherent wholes. The books call the smaller, more concrete ideas that are associated with each big idea *essential understandings*. They capture aspects of the corresponding big idea and provide evidence of its richness.

The big ideas have tremendous value in mathematics. You can gain an appreciation of the power and worth of these densely packed statements through persistent work with the interrelated essential understandings. Grasping these multiple smaller concepts and through them gaining access to the big ideas can greatly increase your intellectual assets and classroom possibilities.

In your work with mathematical ideas in your role as a teacher, you have probably observed that the essential understandings are often at the heart of the understanding that you need for presenting one of these challenging topics to students. Knowing these ideas very well is critical because they are the mathematical pieces that connect to form each big idea.

How are the books organized?

Every book in the Essential Understanding Series has the same structure:

- The introduction gives an overview, explaining the reasons for the selection of the particular topic and highlighting some of the differences between what teachers and students need to know about it.

Big ideas and essential understandings are identified by icons in the books.

marks a big idea, and

marks an essential understanding.

- Chapter 1 is the heart of the book, identifying and examining the big ideas and related essential understandings.

- Chapter 2 reconsiders the ideas discussed in chapter 1 in light of their connections with mathematical ideas within the grade band and with other mathematics that the students have encountered earlier or will encounter later in their study of mathematics.

- Chapter 3 wraps up the discussion by considering the challenges that students often face in grasping the necessary concepts related to the topic under discussion. It analyzes the development of their thinking and offers guidance for presenting ideas to them and assessing their understanding.

The discussion of big ideas and essential understandings in chapter 1 is interspersed with questions labeled "Reflect." It is important to pause in your reading to think about these on your own or discuss them with your colleagues. By engaging with the material in this way, you can make the experience of reading the book participatory, interactive, and dynamic.

Reflect questions can also serve as topics of conversation among local groups of teachers or teachers connected electronically in school districts or even between states. Thus, the Reflect items can extend the possibilities for using the books as tools for formal or informal experiences for in-service and preservice teachers, individually or in groups, in or beyond college or university classes.

marks a "Reflect" question that appears on a different page.

A new perspective

The Essential Understanding Series thus is intended to support you in gaining a deep and broad understanding of mathematics that can benefit your students in many ways. Considering connections between the mathematics under discussion and other mathematics that students encounter earlier and later in the curriculum gives the books unusual depth as well as insight into vertical articulation in school mathematics.

The series appears against the backdrop of *Principles and Standards for School Mathematics* (NCTM 2000), *Curriculum Focal Points for Prekindergarten through Grade 8 Mathematics: A Quest for Coherence* (NCTM 2006), *Focus in High School Mathematics: Reasoning and Sense Making* (NCTM 2009), and the Navigations Series (NCTM 2001–2009). The new books play an important role, supporting the work of these publications by offering content-based professional development.

The other publications, in turn, can flesh out and enrich the new books. After reading this book, for example, you might select hands-on, Standards-based activities from the Navigations books for your students to use to gain insights into the topics that the Essential Understanding books discuss. If you are teaching students in prekindergarten through grade 8, you might apply your deeper understanding as you present material related to the three focal points that Curriculum Focal Points identifies for instruction at your students' level. Or if you are teaching students in grades 9–12, you might use your understanding to enrich the ways in which you can engage students in mathematical reasoning and sense making as presented in *Focus in High School Mathematics*.

An enriched understanding can give you a fresh perspective and infuse new energy into your teaching. We hope that the understanding that you acquire from reading the book will support your efforts as you help your students grasp the ideas that will ensure their mathematical success.

The authors would like to acknowledge Deborah Schifter for her helpful contributions in the preparation of this book. They would also like to acknowledge thoughtful reactions to an earlier draft from Joanne Rossi Becker, Bárbara M. Brizuela, John Lannin, Christina Nugent, and Walter Seaman.

Introduction

This book focuses on ideas about algebraic thinking. These are ideas that you need to understand thoroughly and be able to use flexibly to be highly effective in your teaching of mathematics in grades 3–5. The book discusses many mathematical ideas that are common in elementary school curricula, and it assumes that you have had a variety of mathematics experiences that have motivated you to delve into—and move beyond—the mathematics that you expect your students to learn.

The book is designed to engage you with these ideas, helping you to develop an understanding that will guide you in planning and implementing lessons and assessing your students' learning in ways that reflect the full complexity of algebraic thinking. A deep, rich understanding of the breadth of algebraic thinking will enable you to communicate its influence and scope to your students, showing them how this kind of thinking permeates the mathematics that they have encountered—and will continue to encounter—throughout their school mathematics experiences.

The understanding of algebraic thinking that you gain from this focused study thus supports the vision of *Principles and Standards for School Mathematics* (NCTM 2000): "Imagine a classroom, a school, or a school district where all students have access to high-quality, engaging mathematics instruction" (p. 3). This vision depends on classroom teachers who "are continually growing as professionals" (p. 3) and routinely engage their students in meaningful experiences that help them learn mathematics with understanding.

Why Algebraic Thinking?

Like the topics of all the volumes in NCTM's Essential Understanding Series, algebraic thinking is a major area of school mathematics that is crucial for students to learn but challenging for teachers to teach. Students in grades 3–5 need opportunities to think algebraically—that is, to generalize, express, and justify relationships among quantities, as well as reason with generalizations expressed through a variety of representations, if they are to succeed in their subsequent mathematics experiences. Learners often struggle with aspects of algebraic thinking, not because they are incapable of it, but because they need frequent experiences and time to develop important thinking skills. For example, expressing the solution to an equation such as $37 \times 18 = \_ \times 37$ is important mathematical work, but so are recognizing and using the structure of the equation to reason about the solution in terms of the commutative property of multiplication. Moreover, in tasks like these, many students understand the equals sign only as a signal to perform a computation, without understanding it as a sign of the equivalence of two quantities. The importance of understanding how computation can provide a context for algebraic thinking and the challenge of understanding how learners generalize fundamental properties of operations and develop an understanding of equivalence make it essential for teachers of grades 3–5 to understand these particular aspects of algebraic thinking extremely well themselves.

Your work as a teacher of mathematics in these grades calls for a solid understanding of the mathematics that you—and your school, your district, and your state curriculum—expect your students to learn about algebraic thinking. Your work also requires you to know how this mathematics relates to other mathematical ideas that your students will encounter in the lesson at hand, the current school year, and beyond. Rich mathematical understanding guides teachers' decisions in much of their work, such as choosing tasks for a lesson, posing questions, selecting materials, ordering topics and ideas over time, assessing the quality of students' work, and devising ways to challenge and support their thinking.

Understanding Algebraic Thinking

Teachers teach mathematics because they want others to understand it in ways that will contribute to success and satisfaction in school, work, and life. Helping your students develop a robust and lasting understanding of algebraic thinking requires that you understand this mathematics deeply. But what does this mean?

It is easy to think that understanding an area of mathematics, such as algebraic thinking, means knowing certain facts, being able to solve particular types of problems, and mastering relevant vocabulary. For example, for the upper elementary grades, you are expected to know such facts as, "The product of any whole number and 1 is that whole number." You are expected to be skillful in solving problems that involve equations with one unknown or in computing with large numbers or numbers expressed in fractional or decimal forms. Your mathematical vocabulary is assumed to include such terms as *additive identity, commutative property, equation, function*, and *solution.*

Obviously, facts, vocabulary, and techniques for solving certain types of problems are not all that you are expected to know about algebraic thinking. For example, in your ongoing work with students, you have undoubtedly discovered that you need not only to know common algorithms for addition, subtraction, multiplication, and division, but also to be able to understand, evaluate, and justify strategies that your students create.

It is also easy to focus on a very long list of mathematical ideas that all teachers of mathematics in grades 3–5 are expected to know and teach about algebraic thinking. Curriculum developers often devise and publish such lists. However important the individual items might be, these lists cannot capture the essence of a rich understanding of the topic. Understanding algebraic thinking deeply requires you not only to know important mathematical ideas but also to recognize how these ideas relate to one another. Your understanding continues to grow with experience and as a result of opportunities to embrace new ideas and find new connections among familiar ones.

Furthermore, your understanding of algebraic thinking should transcend the content intended for your students. Some of the differences between what you need to know and what you expect students to learn are easy to point out. For instance, your understanding of the topic should include a grasp of independent and dependent variables and different types of functions—mathematics that students will encounter later but do not yet understand.

Other differences between the understanding that you need to have and the understanding that you expect your students to acquire are less obvious, but your experiences in the classroom have undoubtedly made you aware of them at some level. For example, how many times have you been grateful to have an understanding of algebraic thinking that enables you to recognize the merit in a student's unanticipated generalization or claim, such as, "When I add two odd numbers, I always get an even number?" How many other times have you wondered whether you could be missing such an opportunity or failing to use it to full advantage because of a gap in your knowledge?

As you have almost certainly discovered, knowing and being able to do familiar mathematics are not enough when you're in the classroom. You also need to be able to identify and justify or refute novel claims. These claims and justifications might draw on ideas or techniques that are beyond the mathematical experiences of your students and current curricular expectations for them. For example, you may need to be able to refute erroneous claims such as, "You can't subtract a larger number from a smaller number, so 5 – 8 has to be 3." Or you may need to explain to a student using f to represent the number of feet in the length of a field and y to represent the number of yards in that length, why $y = 3f$ is false, despite the fact that people say, "1 yard is 3 feet."

Big Ideas and Essential Understandings

Thinking about the many particular ideas that are part of a rich understanding of algebraic thinking can be an overwhelming task. Articulating all of those mathematical ideas and their connections would require many books. To choose which ideas to include in this book, the authors considered a critical question: What is *essential* for teachers of mathematics in grades 3–5 to know about algebraic thinking to be effective in the classroom? To answer this question, the authors drew on a variety of resources, including research on mathematics learning and teaching, the expertise of colleagues in mathematics and mathematics education, the reactions of reviewers and professional development providers, ideas from curricular materials, and even personal experiences.

As a result, the mathematical content of this book focuses on essential ideas for teachers about algebraic thinking. In particular, chapter 1 is organized around five big ideas related to this important area of mathematics. Each big idea is supported by smaller, more specific mathematical ideas, which the book calls *essential understandings*.

Benefits for Teaching, Learning, and Assessing

Understanding algebraic thinking can help you implement the Teaching Principle enunciated in *Principles and Standards for School Mathematics*. This Principle sets a high standard for instruction: "Effective mathematics teaching requires understanding what students know and need to learn and then challenging and supporting them to learn it well" (NCTM 2000, p. 16). As in teaching about other critical topics in mathematics, teaching about algebraic thinking requires knowledge that goes "beyond what most teachers experience in standard preservice mathematics courses" (p. 17).

Chapter 1 comes into play at this point, offering an overview of algebraic thinking that is intended to be more focused and comprehensive than many discussions of the topic that you are likely to have encountered. This chapter enumerates, expands on, and gives examples of the big ideas and essential understandings related to algebraic thinking, with the goal of supplementing or reinforcing your understanding. Thus, chapter 1 aims to prepare you to implement the Teaching Principle fully as you provide the support and challenge that your students need to develop robust algebraic thinking.

Consolidating your understanding in this way also prepares you to implement the Learning Principle outlined in *Principles and*

Standards: "Students must learn mathematics with understanding, actively building new knowledge from experience and prior knowledge" (NCTM 2000, p. 20). To support your efforts to help your students learn about algebraic thinking in this way, chapter 2 builds on the understanding of algebraic thinking that chapter 1 communicates by pointing out specific ways in which the big ideas and essential understandings connect with mathematics that students typically encounter earlier or later in school. This chapter supports the Learning Principle by emphasizing longitudinal connections in students' learning about algebraic thinking. For example, as their mathematical experiences expand, students gradually develop an understanding of connections between arithmetic and algebra and become fluent in reasoning with generalizations about numbers and operations.

The understanding that chapters 1 and 2 convey can strengthen another critical area of teaching. Chapter 3 builds on the first two chapters to show how an understanding of algebraic thinking can help you select and develop appropriate tasks, techniques, and tools for assessing your students' understanding of variables, the equals sign, algebraic ideas embedded in arithmetic, functions, equations, and inequalities. An ownership of the big ideas and essential understandings related to algebraic thinking, reinforced by an understanding of students' past and future experiences with the ideas, can help you ensure that your classroom practice reflects the Process Standards and supports the learning of significant mathematics.

Such assessment satisfies the first requirement of the Assessment Principle set out in *Principles and Standards*: "Assessment should support the learning of important mathematics and furnish useful information to both teachers and students" (NCTM 2000, p. 22). An understanding of algebraic thinking can also help you satisfy the second requirement of the Assessment Principle, by enabling you to develop assessment tasks that give you specific information about what your students are thinking and what they understand. For example, a simple task such as $3 + 4 = \_\_\_ + 5$ can reveal important information about children's understanding of equations and the use of symbols to represent equivalent quantities. The student's response to the question, "Is $3 + 4 = 7 + 5$ a true equation?" can reveal much about how the student understands the equals sign and equations, beyond the use of familiar symbols to express computations.

Ready to Begin

This introduction has painted the background, preparing you for the big ideas and associated essential understandings related to algebraic thinking that you will encounter and explore in chapter 1. Reading the chapters in the order in which they appear can be a very useful way to approach the book. Read chapter 1 in more than one sitting, allowing time for reflection. Absorb the ideas—both big ideas and essential understandings—related to algebraic thinking. Appreciate the connections among these ideas. Carry your new-found or reinforced understanding to chapter 2, which guides you in seeing how the ideas in chapter 1 are connected to the mathematics that your students have encountered earlier or will encounter later in school. Then read about teaching, learning, and assessment issues in chapter 3.

Alternatively, you may want to take a look at chapter 3 before engaging with the mathematical ideas in chapters 1 and 2. Having the challenges of teaching, learning, and assessment issues clearly in mind, along with possible approaches to them, can give you a different perspective on the material in the earlier chapters.

No matter how you read the book, let it serve as a tool to expand your understanding, application, and enjoyment of thinking algebraically.

Early Algebra: The Big Ideas and Essential Understandings

How would you answer the question, "What is the essential mathematics content that I need to know to prepare my students for success in elementary grades and beyond?" Your first response might not involve algebra. Yet, in recent years, we have come to view algebra in the elementary grades as essential to helping young children become mathematically successful in school in later years and, ultimately, giving them access to a wide array of careers in technical fields that are critical in the twenty-first century. Consider the question in Reflect 1.1.

> ### Reflect 1.1
>
> How would you define *algebra*? Make a list of what you think might be its essential components.

Historically, the general approach to teaching mathematics treated algebra as an isolated topic for high school, or perhaps middle school. Elementary teachers did not teach it, and presumably did not need to know it. But students' difficulties and resulting high failure rates in typical high school algebra courses (Kaput 2008; Kilpatrick, Swafford, and Findell 2001), coupled with algebra's gatekeeping role in school mathematics, led to significant questions about this traditional treatment of algebra. What emerged was the argument that integrating algebra across prekindergarten through grade 12 could provide the coherence and depth that were lacking. Advocating this approach to algebra, Kilpatrick, Swafford, and Findell (2001) explained:

> The study of algebra need not begin with a formal course in the subject. Recent research and development efforts have been

encouraging. By focusing on ways to use the elementary and middle school curriculum to support the development of algebraic reasoning, these efforts attempt to avoid the difficulties many students now experience and to lay a better foundation for secondary school mathematics. (p. 280)

In conjunction with this shift in perspective on algebra, *Principles and Standards for School Mathematics* (National Council for Teachers of Mathematics [NCTM] 2000) provided a road map for a longitudinal approach to teaching and learning algebra. *Curriculum Focal Points for Prekindergarten through Grade 8 Mathematics: A Quest for Coherence* (NCTM 2006) and the Common Core State Standards for Mathematics (Common Core State Standards Initiative 2010) more recently reiterated this approach by connecting algebra to the critical ideas that young children need to learn. As a result of these and other efforts, core ideas of algebra have now become an essential part of the knowledge that elementary teachers need to teach mathematics.

Characterizing Early Algebra

What do we mean by *algebra* in the elementary grades, or *early algebra*? How is it like—or different from—its counterpart in secondary grades?

Perhaps one of the most significant distinguishing features of early algebra is that it does not resemble the type of algebra that you are likely to have studied in a high school algebra course. It does not focus on the transformational aspects of algebra—that is, techniques or procedures for solving equations or simplifying expressions. Thus, although formal algebra has its historical roots "in the study of general methods for solving equations" (NCTM 2000, p. 37), early algebra brings a more eclectic perspective to the kinds of activities that we might describe as algebra. As we will see, it offers multiple points of entry that draw on arithmetic, functional thinking, mathematical modeling, and quantitative reasoning (Carraher and Schliemann 2007).

Broadly speaking, the heart of early algebra is in generalizing mathematical ideas, representing and justifying generalizations in multiple ways, and reasoning with generalizations (Kaput 2008). Like algebra at any level, as described in NCTM's *Guiding Principles for Mathematics Curriculum and Assessment* (2009), early algebra is a way to "explore, analyze, and represent mathematical concepts and ideas ... and to generalize mathematical ideas and relationships, which apply to a wide variety of mathematical and nonmathematical settings" (p. 4). For example, consider the setting in Reflect 1.2.

In answering the question posed in Reflect 1.2, you engaged in a mathematical activity described as generalizing. In particular,

In this book, examples with children are taken from classrooms across all elementary grades, and all names are pseudonyms.

For an discussion of generalizing and its role in mathematical reasoning in the years leading to high school mathematics, see *Developing Essential Understanding of Reasoning for Teaching Mathematics in Prekindergarten– Grade 8* (Lannin, Ellis, and Elliott forthcoming).

See Reflect 1.2 on p. 39.

Reflect 1.2

Chair and Leg Problem

Suppose that you have some chairs, and each chair has 4 legs. How would you describe the relationship between the number of chairs and the corresponding number of chair legs?

the task required you to identify a general relationship between an unknown but varying number of chairs and the corresponding number of chair legs. We might describe this relationship in a statement such as, "The number of chair legs is four times the number of chairs." It is a *general* relationship because it characterizes how these two quantities relate for any number of chairs. Generalizing is the process by which we identify structure and relationships in mathematical situations. As in the preceding example, it can refer to identifying relationships between quantities that vary in relation to each other. It can also mean lifting out and expressing arithmetic structure in operations on the basis of repeated, regular observations of how these operations behave. For example, a child learning to add whole numbers might notice that the order in which two numbers are added does not matter—that is, the result of the computation will be the same, regardless of the order. Depending on the task and the types of numbers, the child might also form other generalizations, such as, "Any time you add an even number and an odd number, the result is an odd number."

Moreover, generalizing is never distinct from the language by which the *result* of this activity—a generalization—is expressed or represented. Generalizations may be expressed in a number of ways—through natural language, through algebraic notation using letters as variables (discussed in detail in relation to Big Idea 3), or even through tables and graphs. For example, the commutative property of addition might be expressed in the statement, "You can add two numbers in either order" or, more formally, as $a + b = b + a$, where a and b represent any real numbers.

Note that in the Chair and Leg problem in Reflect 1.2, the relationship between the number of chairs and the number of chair legs might be represented as "The number of chair legs is four times the number of chairs." It could also be expressed symbolically, as $l = 4 \times c$, or simply as $l = 4c$, where c represents the number of chairs and l represents the number of legs. It could also be represented in a function table or graph (see fig. 1.1).

In addition to generalizing and expressing generalizations, early algebra involves extending one's thinking beyond producing a generalization to reasoning with generalizations as objects themselves. In this sense, algebraic thinking includes reasoning

This book uses literal symbols (letters) to represent variables. Some researchers have found that the use of non-literal symbols, such as Δ, in early algebra can lead to misconceptions. Researchers have also found that most first-grade children are able to use letters as variables.

Big Idea 3

Variables are versatile tools that are used to describe mathematical ideas in succinct ways.

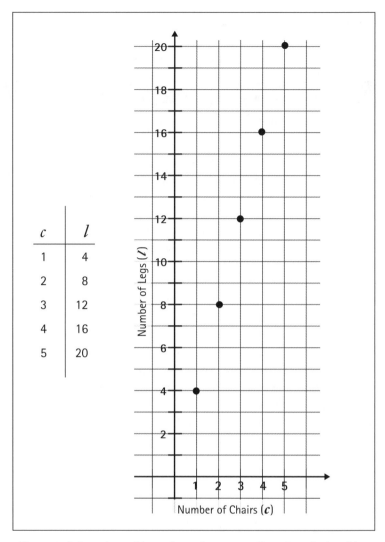

Fig. 1.1. A function table and graph representing the relationship between the number of chairs and the number of chair legs

with structural forms without the need to call particular numbers into play. For example, one child reasoned that $a + b - b = a$ is true because $b - b = 0$ and $a + 0 = a$ (Carpenter, Franke, and Levi 2003). In particular, rather than consider specific computations in the form $a + b - b = a$ (such as $3 + 7 - 7 = 3$), she reasoned with the generalizations $b - b = 0$ and $a + 0 = a$ to argue that $a + b - b = a$ was true. One group of children reasoned that the sum of any three odd numbers is odd by using generalizations that the children had previously established—namely, that the sum of any two odd numbers is even and the sum of any even number and any odd number is odd (Blanton and Kaput 2005). Their argument, "The sum would have to be odd because two odds make an even, and when you add odd plus even, you get odd," was not based on adding sums of three specific odd numbers, but on reasoning with generalizations as objects themselves.

Identifying Algebra Content for Elementary Teachers

NCTM's *Principles and Standards for School Mathematics* (2000, p. 37) identifies four organizing practices in its algebra strand for prekindergarten through grade 12. All students should do the following:

- Understand patterns, relations, and functions
- Represent and analyze mathematical situations and structures using algebraic symbols
- Use mathematical models to represent and understand quantitative relationships
- Analyze change in various contexts

These principles help to detail what we mean by algebra and the particular forms that it might take. Moreover, recent research (see, for example, Kaput, Carraher, and Blanton [2008]) allows us to be more ambitious in this book in defining particular content for elementary teachers.

We have learned that children can think mathematically—algebraically—in even more powerful ways than were envisioned a decade ago. This understanding changes what elementary teachers need to know. As a result, our goal in chapter 1 is to identify essential algebra content that *elementary teachers* need to know, on the basis of our current understanding of how young children reason algebraically. Although some of the ideas discussed in this chapter are those that we would expect children to understand, not all of them are. Where appropriate, we make some of these distinctions explicit. However, it is critical that elementary teachers have a deeper knowledge of particular mathematical ideas so that they can guide students' thinking in appropriate ways and understand how these ideas connect across grades outside their own grade band.

We emphasize that chapter 1 is about algebra content for elementary teachers. We do not intend to develop ideas about how children understand or might learn this content, nor how teachers might teach it. Although we do address some of these ideas briefly in chapters 2 and 3, we encourage you to explore the rich early algebra research base for detailed classroom illustrations about teaching and learning early algebra.

The Big Ideas and Essential Understandings

We do not intend for you to view the big ideas of chapter 1 as mutually exclusive, although we have necessarily presented them that way. Instead, they represent an effort to consolidate core algebra ideas for teachers. The chapter organizes this core knowledge around five big ideas related to different areas: (1) arithmetic as a context for algebraic thinking; (2) equations; (3) variables; (4) quantitative reasoning; and (5) functional thinking. In the remainder of this chapter, we look in more detail at the algebra in each of these areas, and we close the chapter with a brief look at how the big ideas are connected.

Each of the five big ideas that organize this chapter's discussion involves several smaller, more specific "essential understandings." The big ideas and all the associated understandings are identified as a group below to give you a quick overview and for your convenience in referring back to them later. Read through them now, but do not think that you must absorb them fully at this point. The chapter will discuss each one in turn in detail.

Big Idea 1. Addition, subtraction, multiplication, and division operate under the same properties in algebra as they do in arithmetic.

> **Essential Understanding 1a.** The fundamental properties of number and operations govern how operations behave and relate to one another.

> **Essential Understanding 1b.** The fundamental properties are essential to computation.

> **Essential Understanding 1c.** The fundamental properties are used more explicitly in some computation strategies than in others.

> **Essential Understanding 1d.** Simplifying algebraic expressions entails decomposing quantities in insightful ways.

> **Essential Understanding 1e.** Generalizations in arithmetic can be derived from the fundamental properties.

Big Idea 2. A mathematical statement that uses an equals sign to show that two quantities are equivalent is called an *equation*.

> **Essential Understanding 2a.** The equals sign is a symbol that represents a relationship of equivalence.

> **Essential Understanding 2b.** Equations can be reasoned about in their entirety rather than as a series of computations to execute.

Essential Understanding 2c. Equations can be used to represent problem situations.

Big Idea 3. Variables are versatile tools that are used to describe mathematical ideas in succinct ways.

Essential Understanding 3a. The meaning of *variable* can be interpreted in many ways.

Essential Understanding 3b. A variable represents the measure or amount of an object, not the object itself.

Essential Understanding 3c. The same variable used more than once in the same equation must represent identical values in all instances, but different variables may represent the same value.

Essential Understanding 3d. The same variable may play one or more roles within a given application, problem, or situation.

Essential Understanding 3e. A variable may represent either a discrete or a continuous quantity.

Big Idea 4. Quantitative reasoning extends relationships between and among quantities to describe and generalize relationships among these quantities.

Essential Understanding 4a. Two quantities can relate to each other in one of three ways: (1) they can be equal, (2) one quantity can be larger than the other, or (3) one quantity can be smaller than the other.

Essential Understanding 4b. Known relationships between two quantities can be used as a basis for describing relationships with other quantities.

Big Idea 5. Functional thinking includes generalizing relationships between covarying quantities, expressing those relationships in words, symbols, tables, or graphs, and reasoning with these various representations to analyze function behavior.

Essential Understanding 5a. A function is a special mathematical relationship between two sets, where each element from one set, called the *domain*, is related uniquely to an element of the second set, called the *co-domain*.

Essential Understanding 5b. Functions can be viewed as tools for expressing covariation between two quantities.

Essential Understanding 5c. In a functional relationship between two covarying quantities, a variable is said to be either *independent* or *dependent* and will represent either a discrete or a continuous quantity.

Essential Understanding 5d. In working with functions, several important types of patterns or relationships might be observed among quantities that vary in relation to each other: recursive patterns, covariational relationships, and correspondence rules.

Essential Understanding 5e. Functions can be represented in a variety of forms, including words, symbols, tables, and graphs.

Essential Understanding 5f. Different types of functions behave in fundamentally different ways, and analyzing change, or variation, in function behavior is one way to capture this difference.

Arithmetic as a Context for Algebraic Thinking: Big Idea 1

Big Idea 1. *Addition, subtraction, multiplication, and division operate under the same properties in algebra as they do in arithmetic.*

Traditionally, arithmetic has focused on computational accuracy and efficiency. In contrast, much of algebraic problem solving focuses on reasoning about operations. It is possible and highly productive to include reasoning about operations in arithmetic instruction. When reasoning about operations, one learns general properties that describe how operations work. In grades 3–5, understanding these general properties is an essential aspect of algebra content that enhances the learning of both arithmetic and algebra.

The Common Core State Standards for Mathematics so strongly value integrating reasoning about operations into arithmetic instruction that these standards identify "Operations and Algebraic Thinking" as a single domain in kindergarten through grade 5 (Common Core State Standards Initiative 2010). The algebraic thinking described in specific standards in the Operations and Algebraic Thinking domain falls under our Big Idea 1.

The fundamental properties

Essential Understanding 1a. *The fundamental properties of number and operations govern how operations behave and relate to one another.*

The actions of the arithmetic operations are determined and interrelated according to the fundamental properties shown in table 1.1.

Two observations are important to make about table 1.1. First, the operations of subtraction and division are notably absent. Subtraction is defined in terms of its inverse relationship with addition. One way to characterize this relationship is to say that if $a + b = c$, then $c - b = a$. For example, since $10 + (-2) = 8$, then $8 - (-2) = 10$. Similarly, division is defined in terms of its inverse relationship with multiplication. This relationship might be described symbolically by saying that if $a \times b = c$ and $b \neq 0$, then $c \div b = a$. For example, since $16 \times {}^3/_4 = 12$, then $12 \div {}^3/_4 = 16$. The inverse relationship between addition and subtraction (or multiplication and division) is not only foundational to algebra but also deeply arithmetic.

For an extended discussion of the inverse relationship between addition and subtraction, see *Developing Essential Understanding of Addition and Subtraction for Teaching Mathematics in Prekindergarten–Grade 2* (Caldwell, Karp, and Bay-Williams 2011).

For a discussion of the inverse relationship between multiplication and division, see *Developing Essential Understanding of Multiplication and Division for Teaching Mathematics in Grades 3–5* (Otto et al. 2011).

Table 1.1

Fundamental Properties of Number and Operations

Name of property	Representation of property	Example of property, using real numbers
Properties of addition		
Associative	$(a + b) + c = a + (b + c)$	$(78 + 25) + 75 = 78 + (25 + 75)$
Commutative	$a + b = b + a$	$2 + 98 = 98 + 2$
Additive identity	$a + 0 = a$ and $0 + a = a$	$9875 + 0 = 9875$
Additive inverse	For every real number a, there is a real number $-a$ such that $a + -a = -a + a = 0$.	$-47 + 47 = 0$
Properties of multiplication		
Associative	$(a \times b) \times c = a \times (b \times c)$	$(32 \times 5) \times 2 = 32 \times (5 \times 2)$
Commutative	$a \times b = b \times a$	$10 \times 38 = 38 \times 10$
Multiplicative identity	$a \times 1 = a$ and $1 \times a = a$	$387 \times 1 = 387$
Multiplicative inverse	For every real number a, $a \neq 0$, there is a real number $\frac{1}{a}$ such that $a \times \frac{1}{a} = \frac{1}{a} \times a = 1.$	$\frac{8}{3} \times \frac{3}{8} = 1$
Distributive property of multiplication over addition		
Distributive	$a \times (b + c) = a \times b + a \times c$	$7 \times (50 + 2) = 7 \times 50 + 7 \times 2$

(Variables a, b, and c represent real numbers.)

For an extended discussion of rational numbers and their relationship to whole numbers, see *Developing Essential Understanding of Rational Numbers for Teaching Mathematics in Grades 3–5* (Barnett-Clarke et al. 2010).

The second observation that is important to make about table 1.1 is that the fundamental properties of number and operations are true for all real numbers. The set of real numbers consists of two types of numbers: rational numbers and irrational numbers. Most of the numbers encountered in grades 3–5 are rational numbers. A rational number is any number that can be expressed as a/b, where a and b are integers and b is not equal to zero. Integers are formally defined as the set of whole numbers (including zero) and the additive inverses of the whole numbers (often thought of as the negative whole numbers). All of the numbers that fit in the sequence $\{\dots, -3, -2, -1, 0, 1, 2, 3, \dots\}$ are integers. Rational numbers include whole numbers, since a whole number can be expressed as itself divided by 1. For example, 5 can be expressed as $5/1$. On the other hand, irrational numbers are those real numbers that are not rational. In particular, they are numbers that can be written as non-terminating, non-repeating decimals. Well-known examples of irrational numbers include $\sqrt{2}$ and π. Unlike rational numbers, irrational numbers cannot be expressed as the quotient of two integers.

Algebra and the fundamental properties

Essential Understanding 1b. *The fundamental properties are essential to computation.*

The fundamental properties listed in table 1.1 are the relationships that govern how operations work in arithmetic and algebra. (Other relationships that are related to the fundamental properties are discussed later in this chapter.) When problem solvers draw on these relationships in solving a problem, they are using what are described as relational thinking strategies (Carpenter, Franke, and Levi 2003; Empson and Levi 2011). The most basic relational thinking strategies involve the intuitive *use of a property to solve a problem.* Consider how Elena solved the following problem:

> Martinique has 6 boxes of marbles, with 23 marbles in each box. How many marbles does Martinique have altogether?

Elena solved the problem as follows:

> I will do the 20s and then the 3s. Six groups of 20 are 120; 6 groups of 3 are 18. There would be 120 plus 18, or 138 marbles.

Although Elena did not provide the following representation, the reasoning behind her solution can be represented with the equation

$$6 \times 23 = (6 \times 20) + (6 \times 3).$$

Elena used an intuitive understanding of the distributive property of multiplication over addition (which we will hereafter shorten to "the distributive property") to solve this problem. When asked why she could solve the problem like this, Elena replied, "It would be easier to just do the 20s and then do the 3s. I know I have all of the marbles, since there are 6 boxes with 23 in each."

Elena's relational thinking strategy illustrates a preliminary understanding of the distributive property. Not only is this type of knowledge essential for a deep understanding of arithmetic, it is also an important starting point for developing a more generalized (algebraic) understanding of the distributive property. A more complex way to engage with the fundamental properties is to *make a general statement, or generalization, about a property*. A few months later, Elena used the distributive property to solve a problem about 8 groups of 35. When asked why she solved the problem as she did, Elena said, "Whenever you multiply, you can always break up one of the groups into two parts and then multiply one part and then the other part and then add them back together." We can call this statement a generalization because Elena describes a general process for multiplying that does not depend on the specific numbers used in the problem. Reflect 1.3 takes this thinking a step further.

Reflect 1.3

How would you use an equation to describe Elena's general statement of the distributive property?

For an elaboration of the relationship between generalization and conjecture, see *Developing Essential Understanding of Reasoning for Teaching Mathematics in Prekindergarten– Grade 8* (Lannin, Ellis, and Elliott, forthcoming).

Discussions of the fundamental properties of number and operations and their role in justifying computational strategies and facilitating mental computations appear in *Developing Essential Understanding of Multiplication and Division for Teaching Mathematics in Grades 3–5* (Otto et al. 2011).

One's generalizations about fundamental properties are often initially expressed in a manner similar to Elena's—that is, informally, using natural language. Although this form of expression is sometimes awkward, it is an important step towards formalizing a general equation using variables. For example, we could express Elena's conjecture symbolically as $a \times (b + c) = (a \times b) + (a \times c)$, where a, b, and c represent real numbers.

This symbolic representation of the distributive property indicates that the property holds true for all real numbers, whereas Elena's generalization may imply that the distributive property holds only for multiplication situations that can easily be thought of as groups of objects. An algebraic understanding of the fundamental properties includes understanding that these properties represent generalizations that apply to all real numbers. It also includes being able to represent them symbolically and to identify their use explicitly in computations.

Essential Understanding 1*b* illustrates the critical depth of algebraic knowledge that is associated with understanding arithmetic, particularly as it relates to the fundamental properties. This knowledge begins at an informal level of reasoning, in which one might invoke a property without an explicit knowledge that it is being used, or how. From this starting point, it extends to a general understanding of the property and how it works, as well as how it can be expressed by using formal algebraic notation (variables and equations) and that it is true for all real numbers. Finally, unlike other arithmetic generalizations that can be proven true or false (see Essential Understanding 1*e*), fundamental properties are statements that we assume to be true without proof. However, it is important for children to understand that these properties are true and to be able to reason with them.

Using computation to deepen algebraic understanding

Essential Understanding 1c. The fundamental properties are used more explicitly in some computation strategies than in others.

Most people are aware of the standard algorithms for computation. However, many people—adults and children alike—rely on intuitive relational thinking strategies for computation in certain situations. Situations in which people may rely on intuitive procedures for computation include the following:

- Figuring out how to split a restaurant bill and how much to leave for a tip

- Determining someone's age by using the year that he or she was born

- Estimating the arrival time on a trip on the basis of the traveling speed and a road sign posting the mileage to the destination

Typically, these relational thinking strategies have not been explicitly taught but are strategies that people generate from their understandings of how number and operations work.

Relational thinking can be used to solve many problems traditionally posed in elementary and middle school. Consider the following problem:

$$15 \times 6\frac{4}{5} = p$$

Armin solved the problem by reasoning in this way:

> Fifteen times $^4/_5$ is going to be easy. Since 5 is a factor of 15, I know the answer will be a whole number, so I am doing that first. Fifteen times 4 is 60, so 15 times $^4/_5$ is $^{60}/_5$, which is the same as 12. Now all I have to do is 15 times 6, which is 90, and add it to 12. So, p is 90 plus 12, which is 102.

Armin's solution can be represented with formal mathematical notation as follows:

$$15 \times 6\frac{4}{5} = 15 \times (\frac{4}{5} + 6)$$

$$= (15 \times \frac{4}{5}) + (15 \times 6)$$

$$= \frac{60}{5} + (15 \times 6)$$

$$= 12 + 90$$

$$= 102$$

Armin's solution is based on the distributive property, which is the basis for his reasoning behind the following segment of his solution:

$$15 \times \left(\frac{4}{5} + 6 \right) = \left(15 \times \frac{4}{5} \right) + (15 \times 6)$$

Standard algorithms were developed over hundreds of years to provide procedures that enable us to compute quickly and accurately. They were developed at a time when there were no machines—or even the idea that someday there would be machines—to perform basic computations. Although all standard algorithms are based on the fundamental properties, often these properties are buried deeply in the algorithm. In contrast, the fundamental properties are typically easy to identify in relational thinking strategies, such as Armin's and Elena's, since people generate these procedures on the basis of their understanding of how operations work. Computations can serve as a context for algebraic thinking when one understands and can explicitly identify how the fundamental properties are used in computations. Although students might not explicitly identify the use of these properties as they compute, it is important that teachers understand how these properties are used in computations and how to explicitly identify them. Teachers need a deep algebraic understanding of the properties to help children reflect on their use of them in computation.

Decomposing quantities as a foundation for algebra

Essential Understanding 1d. *Simplifying algebraic expressions entails decomposing quantities in insightful ways.*

Elena's and Armin's relational thinking strategies not only rely on a strong understanding of multiplication, they also rely on being able to decompose numbers in insightful ways. For example, when solving the problem involving 6 groups of 23, Elena could have broken 23 into 16 and 7, determined the amounts in 6 groups of 16 and 6 groups of 7, then added the products together to determine the product of 6 and 23. Even though this would be a valid use of the distributive property, it would not have been an *insightful* way to decompose 23 to solve the problem, since the resulting computations would not be any simpler than the original one.

To use the fundamental properties efficiently during computation, one needs to analyze the problem to determine both what properties might apply and how best to decompose the numbers in the problem to apply these properties effectively. For example, to compute 98 + 55, it makes sense to decompose 55 into 2 + 53, so that 98 + 55 can be expressed as 98 + 2 + 53. To compute 43 + 55, it makes sense to decompose 55 into 50 + 5 and 43 into 40 + 3 so that

For additional insights into composition and decomposition of number and connections of decomposition to other topics in early childhood and elementary school mathematics, see *Developing Essential Understanding of Number and Numeration for Teaching Mathematics in Prekindergarten– Grade 2* (Dougherty et al. 2010).

43 + 55 can be thought of as (40 + 50) + (3 + 5). To compute $55 \times \frac{1}{5}$, it makes sense to decompose 55 into the factors 11 and 5, since

$$55 \times \frac{1}{5} = (11 \times 5) \times \frac{1}{5} = 11 \times (5 \times \frac{1}{5}) = 11 \times 1 = 11.$$

In more formal algebra beyond elementary grades, students often need to determine insightful ways to decompose general quantities. For example, different decompositions of the quantity $4x^2 - 4x$ would be used to simplify the following expressions:

$$\frac{4x^2 - 4x}{x}, \ x \neq 0, \ \text{and} \ \frac{4x^2 - 4x}{x-1}, \ x \neq 1$$

In the first, it makes sense to decompose $4x^2 - 4x$ into $x(4x - 4)$ so that the expression can be simplified to $4x - 4$, $x \neq 0$. In the second, it makes sense to decompose $4x^2 - 4x$ into $4x(x - 1)$ so that the expression can be simplified to $4x$, $x \neq 1$. Learning insightful ways to decompose numbers in arithmetic focuses one's attention on relationships in operations on numbers. It also develops a facility with the use of the fundamental properties in decomposing these numbers (Empson, Levi, and Carpenter 2011; Empson and Levi 2011). This attention to structure and how numbers are composed (and hence can be decomposed) provides an important foundation for determining insightful ways to decompose algebraic expressions in the study of algebra beyond the elementary grades.

Other arithmetic opportunities for developing generalizations

Essential Understanding 1e. *Generalizations in arithmetic can be derived from the fundamental properties.*

Generalizations derived from the fundamental properties can provide a rich context for thinking algebraically. Consider how Eve reasoned about a generalization involving adding and subtracting a fixed amount from a given number (see also Carpenter, Franke, and Levi 2003, pp. 97–102): "If you add a number to a given number and then subtract the same number, the given number stays the same." Eve represented the conjecture with the following mathematical notation:

$$a + b - b = a$$

She circled $b - b$ and said, "This is always zero. We already know that any number minus itself is always zero, so I could just write

$$b - b = 0$$
$$a + 0 = a,$$

and we already know that a plus 0 is always a, so $a + b - b = a$ must always be true." Not only does Eve's reasoning about this conjecture draw on the fundamental properties (for example, the additive identity), but the fact that she was able to use generalized ideas or properties as the basis for her justification provides important evidence of her ability to think algebraically.

Note that by operating first on $b - b$ in the expression $a + b - b$, Eve is assuming that $a + b - b = a + (b - b)$. This assumption is true, but whether Eve understands it—or recognizes that she is making it— is unclear. To simplify the expression $92 - 57 - 7$, might Eve operate first on $57 - 7$? Doing so would not be valid, since $92 - 57 - 7 \neq 92 - (57 - 7)$. See Carpenter, Franke, and Levi (2003, pp. 105–21) for more information about ordering multiple operations.

Another type of generalization derived from the fundamental properties is a generalization about a class or classes of numbers. Computations with odd numbers and even numbers—common classes of numbers for students to work with in grades 3–5—can lead to a variety of generalizations, such as the following:

- An odd number plus an odd number is an even number.

- An even number times an even number is an even number.

- An odd number times an even number is an even number.

- An even number plus an odd number is an odd number.

Building generalizations in response to a question such as, "What can you say about the sum of two odd numbers?" mirrors the reasoning of more formal algebra, which often requires looking for generalized patterns.

Once someone has formed a generalization, working to justify it supports algebraic thinking because it focuses attention on the structure that underlies the computation. Consider how Gabriella justified the conjecture, "An odd number plus an odd number is an even number":

I know that any odd number can be written like this:

$$b + b + 1.$$

As long as b is a whole number, b plus b plus 1 will be odd. You can't let b be a fraction like 1/2 or something. Adding two odd numbers together would be written like this,

$$b + b + 1 + d + d + 1.$$

The b's and d's can't be fractions. They have to be whole numbers. I can write this like,

$$b + b + 1 + d + d + 1 = b + b + d + d + 2,$$

since I know that when you add numbers you can add them in any order. Well, b plus b is even, since any number added to itself is even, and d plus d is also even for the same reason.

Two is even just because it is.

Her paper now looked like figure 1.2.

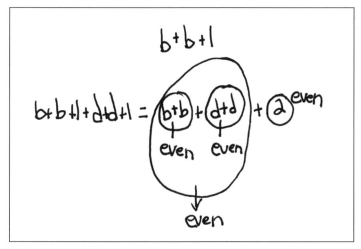

Fig. 1.2. Gabriella's written work justifying the conjecture

Gabriella continued to explain her reasoning:

> So we have an even plus an even—that is just even. We already
> proved that, and then plus another even. Again, that is just
> even. So we showed that any odd number plus another odd
> number is even.

In the preceding examples, Gabriella and Eve invoked and reasoned with generalizations that they already knew were true. They
were also able to express generalized numbers symbolically (for
example, Gabriella expressed an arbitrary odd number as $b + b + 1$)
and operate on these symbolic expressions. This type of reasoning
is not only algebraic, but also foundational to formal mathematical
proof in algebra as well as other areas of mathematics.

If students' early work with arithmetic consists solely of prac-
ticing standard algorithms and memorizing facts to use in those
algorithms, they are unlikely to engage in algebraic thinking during
this study. However, when students' learning of arithmetic focuses
on reasoning about mathematical relationships, they naturally
engage in a great deal of algebraic thinking while thinking about
arithmetic.

A focus on reasoning about mathematical relationships is essential not only for algebra learning but also for developing a deep
understanding of arithmetic. Historically, arithmetic and algebra
were treated as distinct fields of study. Currently, many people see
a great deal of overlap between arithmetic and algebra. Some even
view arithmetic computation and algebraic reasoning about relationships as a single domain. The Common Core State Standards for
Mathematics (Common Core State Standards Initiative 2010) connect

For an elaboration
of relationships
between
generalizations
and justification of
reasoning in
the years before
high school, see
*Developing Essential
Understanding of
Reasoning for Teach-
ing Mathematics in
Prekindergarten–
Grade 8* (Lannin, Ellis,
and Elliott
forthcoming).

the two explicitly, as we observed at the beginning of our discussion of Big Idea 1. Thus, although we have identified reasoning about the fundamental properties as algebraic thinking, a true understanding of arithmetic also includes reasoning about these properties.

Equations as Statements of the Equivalence of Two Quantities: Big Idea 2

Big Idea 2. *A mathematical statement that uses an equals sign to show that two quantities are equivalent is called an equation.*

In the elementary grades, an equation typically represents a relationship of equivalence between two quantities or amounts. Mathematical expressions, which are representations of a single quantity or amount, are often mistaken for equations. The following are examples of expressions:

$$8 \qquad 358 \times 459 \qquad a + 58 \qquad x^2 + 2x + 1$$

In contrast, an equation represents an equivalence relationship between two quantities. The following are examples of equations:

$$8 = 8 \qquad\qquad 358 \times 458 = 458 \times 358$$
$$a + 58 = 59 + 176 \qquad x^2 + 2x + 1 = 0$$

Equations are plentiful in traditional arithmetic instruction. For example, students may be asked to complete equations such as $346 \div 18 =$ ___, or find n in an equation such as $87 \times 42 = n$ or $84 - n = 39$. However, equations such as these are typically used to prompt students to compute rather than to reason about the relationship between the amounts. In grades 3–5, equations can be used for more than prompting computation. They can also be used to develop algebraic thinking.

In an equation, the expressions on the two sides of the equals sign are equivalent, even if the symbols representing those two quantities do not look identical. Using equations to reason about, represent, and communicate relationships between quantities is a cornerstone of algebra. As the following paragraphs describe, equations such as $84 \times 359 = 359 \times j$ and $a + 467 - 467 = 8769$, where the numbers and operations used are those typically found in grades 3–5, can prompt students to reason about the relationships among quantities and, by doing so, develop algebraic understanding.

A relational understanding of the equals sign

Essential Understanding 2a. *The equals sign is a symbol that represents a relationship of equivalence.*

A *relational understanding* of the equals sign entails understanding that the equals sign represents a relationship of equivalence. In the elementary grades, this relationship is often interpreted as meaning "is the same amount as" when expressing a relationship between equivalent amounts (see Carpenter, Franke, and Levi [2003, pp.10–14]). With a relational understanding, one knows that 6 = 6 and 6 = 3 + 3 are true equations and that 8 + 4 = 12 + 5 is false, since 8 + 4 and 12 + 5 are not equivalent. A relational understanding of the equals sign is necessary for the use of relational thinking to solve equations.

To gain a better understanding of what it means to use relational thinking to solve equations, imagine how you might solve a simple equation such as $9 + 6 = b + 7$. Using relational thinking, you might reason that since 7 is 1 more than 6, then b must be 1 less than 9. This ability to reason about structure in the comparison of two quantities is an important aspect of thinking algebraically. As we described earlier, strategies such as this, which draw on a relational understanding of the equals sign, are referred to as relational thinking strategies. As we discuss in the next section, a relational understanding of the equals sign (and the relational thinking strategies that reflect this understanding) is necessary to solve algebraic equations because it allows reasoning about equations in their entirety.

Reasoning about equations

Essential Understanding 2b. *Equations can be reasoned about in their entirety rather than as a series of computations to execute.*

Equations presented in traditional arithmetic instruction are typically thought about as a series of computations to execute. Compare the equations in columns A and B in table 1.2.

Table 1.2
Equations for Comparison

Column A	Column B
$48 \times 67 \times 6 = k$	$347 \times 25 \times 4 = p$
$746 \times 398 \div 42 = t$	$398 \times 746 \div 746 = d$
$978 + 778 = 394 + y$	$378 + 794 = 778 + j$
$475 \times 2365 = 352 \times w$	$8790 \times 598 = 879 \times n$

Even though the equations in column A are structurally identical to the equations in column B, the equations in column A are challenging to solve without enacting a series of computations. Although the equations in column B could be solved by executing a series of computations, they could also be solved by first analyzing the entire equation to look for relationships that might simplify their solution. For example, in the equation $347 \times 25 \times 4 = p$, it is far more efficient to compute 25×4 first to get 100, and then compute

347×100 to obtain $p = 34,700$, than it is to compute in order from the left, beginning with 347×25 and then multiplying by 4. The more efficient solution involves using the associative property of multiplication—one of the fundamental properties—as a way to simplify the indicated product. (See the preceding discussion of Big Idea 1 for more about the fundamental properties.)

Reasoning about equality can also be used to simplify solutions to equations. Consider Tiara's solution to the equation $378 + 794 = 778 + j$: "We have 700 and 78 on both sides of the equation, so you don't have to think about them. You are left with $300 + 94 = j$, so $j = 394$." This solution relies on understanding that if two amounts are equal, you can subtract the same number from each amount, and the results will still be equal. This property is often referred to as the *subtraction property of equality*: if $a = b$, then $a - c = b - c$, for all real numbers $a, b,$ and c. Other properties of equality include the addition property of equality (if $a = b$, then $a + c = b + c$), the multiplication property of equality (if $a = b$, then $a \times c = b \times c$), and the division property of equality (if $a = b$ and $c \neq 0$, then $a \div c = b \div c$).

Relational thinking strategies that draw on these properties of equality almost always involve reasoning about the fundamental properties. Consider again Tiara's solution to the equation $378 + 794 = 778 + j$. The specific steps that justify this solution can be illustrated by the following equations:

$$378 + 794 = 778 + j$$
$$(300 + 78) + (700 + 94) = (700 + 78) + j$$
$$(300 + 78) + 700 + 94 = j + (700 + 78)$$
$$300 + (78 + 700) + 94 = j + (700 + 78)$$
$$300 + 94 + (700 + 78) = j + (700 + 78)$$
$$300 + 94 + (700 + 78) - 700 - 78 = j + (700 + 78) - 700 - 78$$
$$394 + (700 - 700) + (78 - 78) = j + (700 - 700) + (78 - 78)$$
$$394 + 0 + 0 = j + 0 + 0$$
$$394 = j$$

This sequence of equations involves multiple uses of the associative and commutative properties of addition, along with uses of the subtraction property of equality. Sometimes, more than one property is used in the same line. It is not necessary, nor would it be mathematically powerful, to write out all of these steps in solving equations like these. However, these steps are all necessary to justify why this solution works, and an implicit understanding of these properties is essential to the ability to use relational thinking strategies like this to solve equations. Reflect 1.4 asks you to use a relational thinking strategy to solve a different problem.

Big Idea 1

Addition, subtraction, multiplication, and division operate under the same properties in algebra as they do in arithmetic.

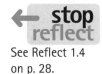

See Reflect 1.4 on p. 28.

Reflect 1.4

How might you use a strategy similar to Tiara's to solve $8790 \times 598 = 879 \times n$?

The equations presented in table 1.2 all contain a single variable and can be readily solved by executing a series of computations. However, if your only way of solving equations relies on executing a series of computations—rather than reasoning about the equation in its entirety—you would be hard-pressed to solve algebraic equations such as $40 + m = 84 - m$ or $x^2 + 10x + 21 = 0$. If you viewed equations as a series of computations to execute, the only strategy that you could use to solve these problems would be trial and error, through substituting values for the variable until you found one that made the equation true. You would not use properties of number and operations or properties of equality to reason about the equation. However, by using these properties to reason about the equation in Reflect 1.4, you can recognize that 8790 is ten times 879, so n needs to be ten times 598, or 5980. Understanding how to solve equations such as these is a basic component of more formal algebra that comes into play in later grades and requires an implicit knowledge of the fundamental properties.

All reasonably efficient strategies for solving equations such as those described in the preceding paragraphs are based on the fundamental properties and properties of equality. Most people are familiar with standard strategies, typically taught in high school algebra, for solving equations. *Nonstandard* strategies for solving equations are particularly relevant to algebra in grades 3–5 because they allow students to reason intuitively about an equation in its entirety. As an example of a nonstandard strategy, consider Kevin's explanation of how he solved the equation $47 - y - y = 30 - y$:

> All of those y's represent the same amount. On one side, you are taking that amount away twice, and on the other side, only once. I am not going to think of taking one of those y's away on the left, and I am not going to think about taking the y away on the right. Since each side is having the same amount subtracted, I don't need to think about it. Then I just have $47 - y = 30$, so y is 17.

➡ **Essential Understanding 4b**
Known relationships between two quantities can be used as a basis for describing relationships with other quantities.

Kevin's strategy is based on the addition property of equality, where y is added to both sides of the equation. Note that his reasoning does not involve a formal application of this property. That is, he does not formally add y to both sides of the equation and then simplify to obtain the resulting equivalent equation. Later, when we discuss Essential Understanding 4b, we will describe how this type of thinking can develop through quantitative reasoning. However, suffice it for now to say that it does entail reasoning about the equation in its entirety in a way that accounts for equivalent quantities on both sides of the equation. Reflect 1.5 asks you to extend Kevin's strategy to solve a new equation.

Reflect 1.5

Using Kevin's solution to the equation $47 - y - y = 30 - y$ as a guide, how might you solve the equation $y + y + y = 30 + y$?

Consider Hidetomi's solution to the equation $40 + m = 84 - m$:

If I add m to 40, I get the same thing as if I take m away from 84. That would be the same as adding m to 40 twice and then getting 84. And 84 is 44 more than 40, so m must be half of 44, or 22.

The steps that justify this solution are illustrated by the sequence of equations below:

$$40 + m = 84 - m$$
$$40 + m + m = 84 - m + m$$
$$40 + 2m = 40 + 44$$
$$2m = 44$$
$$\frac{1}{2} \times 2m = \frac{1}{2} \times 44$$
$$m = 22$$

These equations allow us to see how Hidetomi intuitively used both the addition and the multiplication properties of equality in his solution strategy. Although Hidetomi did not write these equations to solve the problem, they are provided here to illustrate the reasoning behind his solution.

This type of reasoning to solve equations with multiple variables or repeated use of a single variable—the kind of equation that Hidetomi solved—typically involves a more challenging application of the same reasoning used to solve equations that involve only the single use of one variable. Because this is so, reasoning with properties of equality and of number and operations to solve equations with a single variable can provide a foundation for understanding how to solve more complex equations. Not all equations with a single variable are well suited to provide the opportunity for this type of reasoning, but understanding how to recognize those that do allow such reasoning is an indication of the development of an algebraic understanding of equations. Reflect 1.6 probes an application of properties of equality.

See Reflect 1.6 on p. 30.

Reflect 1.6

Using the equation that Hidetomi solved as a guide, how might you use properties of equality to solve $4 \times m = 36 \div m$?

One might think of this equation structurally as "4 times some number m is the same as 36 divided by that same number m." Because 4×3 is 12 and $36 \div 3$ is also 12, $m = 3$ is a solution. (Note that $m = -3$ is also a solution, since $4 \times (-3)$ is -12 and $36 \div (-3)$ is also -12.) Alternatively, one might think of this equation structurally as "4 times some number m, times that number again, is the same as 36." Because $4 \times (3 \times 3)$ is 36, m could be 3. (Similarly, m could be -3, since $4 \times ((-3) \times (-3))$ is 36.)

Using equations to model problem situations

Essential Understanding 2c. *Equations can be used to represent problem situations.*

Many problems are best solved by first writing an equation or equations to represent the situation in the problem. The equations written to represent these situations become the tools that enable one to solve the problem.

Most arithmetic word problems can be solved without writing an equation to represent the problem. Nevertheless, there is almost always an equation or equations that represent the situation in the problem. Writing equations to represent situations in arithmetic word problems can help students develop an understanding of how equations can be written to represent algebra problem situations, thus building a critical foundation for learning more formal algebra in later grades.

Sometimes, the way in which one solves a problem does not match the equation that represents the situation in the problem. For example, consider the following problem:

> JaeQwan is making flowerpots. One flowerpot takes $3/4$ of a pound of clay. How many flowerpots can JaeQwan make with $4 1/2$ pounds of clay?

Jan explained her solution: "Three-fourths plus $3/4$ is $1 1/2$—that is 2 pots. Plus another $3/4$ is $2 1/4$—that is 3 pots. Plus another $3/4$ is 3—that is 4 pots. Plus $3/4$ is $3 3/4$—that is 5 pots. Plus $3/4$ is $4 1/2$—that is 6 pots." The equation that represents how this problem was solved is as follows:

$$\frac{3}{4}+\frac{3}{4}+\frac{3}{4}+\frac{3}{4}+\frac{3}{4}+\frac{3}{4}=4\frac{1}{2}$$

Representing the structure of addition and subtraction problems is discussed in detail in *Developing Essential Understanding of Addition and Subtraction for Teaching Mathematics in Prekindergarten–Grade 2* (Caldwell, Karp, and Bay-Williams 2011).

The equation that represents the situation in the problem could be:

$$4\frac{1}{2} \div \frac{3}{4} = m \quad \text{or} \quad m \times \frac{3}{4} = 4\frac{1}{2}.$$

Writing an equation to represent a problem situation often requires a more generalized understanding of number and operations than writing an equation to represent the way that a problem was solved. Although Jan solved this problem without realizing that it was a problem involving division of fractions, eventually she will come across numbers in fraction division problems for which a solution strategy of repeated addition is too complicated.

For example, what if JaeQwan owned a pottery factory and had 4,327 pounds of clay? Knowing that this is a division problem and knowing how to write a corresponding equation might be necessary to solve it efficiently. A deep understanding of algebra requires solving problems like this, in which either the amount of clay that JaeQwan had or the amount needed for each pot varies, or in which both amounts vary. When the amount for each pot or the total amount of clay varies, an equation is an important tool to use in reasoning about the problem. Writing equations that represent the situation in arithmetic problems builds a foundation for writing equations in algebra.

Detailed discussion of representing the structure of division and multiplication problems is provided in *Developing Essential Understanding of Multiplication and Division for Teaching Mathematics in Grades 3–5* (Otto et al. 2011).

Variables as Versatile Tools: Big Idea 3

Big Idea 3. *Variables are versatile tools that are used to describe mathematical ideas in succinct ways.*

Algebraic ideas can be expressed or verbalized in many ways that do not involve standard algebraic notation. These might include natural language, graphs, or tables. However, an understanding of variable and the ability to use variables to express generalizations are important components of the development of algebraic thinking.

The different meanings of *variable*

Essential Understanding 3a. *The meaning of* variable *can be interpreted in many ways.*

Constructing a definition for *variable* is complicated by the fact that the term can have different meanings, depending on the context in which it is used. In mathematics, *variable* can be used to describe many different mathematical situations:

- A variable can represent a number in a generalized pattern.

  Example: The letters r and s in the equation $r + s = s + r$, where r and s represent real numbers

  While trying to learn basic facts in arithmetic, Emily observed that $2 + 9 = 9 + 2$, $3 + 8 = 8 + 3$, $1 + 7 = 7 + 1$, $2 + 8 = 8 + 2$, $5 + 4 = 4 + 5$, and so forth. She then conjectured that the sum of two numbers remains the same if the order of the two addends is reversed, regardless of the values of the two addends. She wanted to tell her teacher about her conjecture, but she soon realized that she could not write this relationship for all possible sums of two numbers. She began by expressing her conjecture to the class in natural language: "I can add two numbers in any order, and the result is the same." With the teacher's guidance, she was able to symbolize two arbitrary numbers as r and s and use these representations to communicate her generalization efficiently by writing $r + s = s + r$ (which we know as the commutative property of addition). In this sense, Emily used variables to represent a number in a generalized pattern.

- A variable can be used to represent a fixed but unknown number.

  Examples: The letter y in the equation $y + 5 = 8$ and the letter x in the equation $2x - 3 = x + 1$

At first, the values of y and x that make the equations in the examples true are unknown. Furthermore, for the equations to be true, the values of y and x must be fixed: the variables cannot represent any arbitrary value. For example, letting y represent the numbers 0 or 5 does not make the first equation true. Likewise, letting x represent the number 1 or 7 does not make the second equation true. By solving the equations (or maybe just by inspecting them), we can determine that the values of the variables that make their respective equations true are $y = 3$ and $x = 4$. As we saw earlier in the discussion of equations in relation to Big Idea 2, Hidetomi solved an equation where the variable, m, represented a fixed but unknown number that he found to be 22.

- A variable can be used to represent a quantity that varies, especially in relation to another quantity.

Example: The letters y and f in the equation $3y = f$

There are many values of y and f that make the equation in the example true (for instance, $y = 1$ and $f = 3$, $y = 2$ and $f = 6$, or $y = 3$ and $f = 9$). However, the values of y and f do not vary randomly. The value of f is always three times the value of y. For example, $y = 10$ and $f = 5$ would not be values that make the equation true, since 5 is not 3 times 10. When variables are used in this way, it is often the relationship between the two variables that is of interest, and not the values of the variables themselves. (See the discussion of Big Idea 5, related to functional thinking, for more on relationships between two quantities.)

- A variable can be used to represent a parameter.

Example: The letter m in the equation $y = mx$

A parameter can be thought of as a quantity whose value determines the characteristics or behavior of other quantities. In the Chair and Leg problem in Reflect 1.2 (see p. 9), if each chair had 5 legs instead of 4, the relationship could be written as $y = 5x$, where x represents the number of chairs and y represents the number of chair legs. If each chair had 6 legs, the relationship would become $y = 6x$. What happens if we generalize the number of legs on each chair to m legs? If m is the number of legs on each chair, then we can express the relationship between the number of chairs and the number of legs as $y = mx$. The variable m is considered to be a parameter because it determines the behavior of the specific function involving the variables x and y. This example also illustrates that variables can take on different

Big Idea 2

A mathematical statement that uses an equals sign to show that two quantities are equivalent is called an equation.

Big Idea 5

Functional thinking includes generalizing relationships between covarying quantities, expressing those relationships in words, symbols, tables, or graphs, and reasoning with these various representations to analyze function behavior.

➡ Essential
Understanding 3d
The same variable
may play one or
more roles within a
given application,
problem, or
situation.

but essential roles within the same context (see Essential Understanding 3d). In this case, the variable m acts as a parameter that is fixed for each problem situation, while the variables x and y represent quantities that vary within the given problem situation.

- A variable can be used to represent an arbitrary or abstract placeholder in an algebraic process.

Example: The letter t in the statement, "Factor $t^2 + 3t$"

In this example, it is not very useful to think of the variable t as representing a particular number. That is, the expression $t^2 + 3t$ is not an equation to be solved for t. To factor $t^2 + 3t$, the variable t is considered an abstract symbol to be manipulated according to an established set of rules (for example, the distributive property of multiplication over addition). Using these rules, it is permissible to rewrite $t^2 + 3t$ as $t(t + 3)$. (An example of using a variable as a placeholder appears in the discussion of Essential Understanding 1d in relation to decomposing quantities as a foundation for algebra; see p. 21 for different decompositions of the quantity $4x^2 - 4x$ that would be useful in simplifying different expressions.)

➡ Essential
Understanding 1d
Simplifying
algebraic expressions
entails decomposing
quantities in
insightful ways.

Although it is difficult to construct a definition of the term *variable* that is appropriate for all situations, what should be clear is that a variable can mean more than a symbol that represents an unknown number. Reflect 1.7 asks you to explore this notion.

Reflect 1.7

How would you describe the role played by the variable t in each of the following statements?

$t + 4 = 3t - 6$

$y = tx + 2$

$3 + (t + 5) = (3 + t) + 5$

Were you able to distinguish and interpret how each variable is being used in each particular context in Reflect 1.7? In the statements provided, t would usually be thought of as a fixed, but unknown, quantity (in statement 1); a parameter (in statement 2); and a symbol for expressing a generalization (in statement 3). However, you may have come up with different answers. It is not always a simple task to determine the role of a variable, since two people may think differently about a problem and therefore have different perspectives about the role that a particular variable plays.

In most cases, the role that a variable plays is determined by the context or physical situation in which the equation or expression is set. For example, figure 1.3 shows a table that David has made. David represents the numbers in the first row by the variable x and the numbers in the second row by the variable y.

x	1	2	3	4	5	6
y	2	4	6	8	10	12

Fig. 1.3. David's table of data

David is interested in finding other values of x and y that fit this pattern. He is thinking of the variables as varying quantities, and he extends his table as shown in figure 1.4. Reflect 1.8 explores how David might be thinking about x and y.

x	1	2	3	4	5	6	7	8	9	10
y	2	4	6	8	10	12	14	16	18	20

Fig. 1.4. David's extended table of data

Reflect 1.8

Mr. Lewis believes that David is thinking of the variables x and y as generalizing a pattern. What would you say to Mr. Lewis?

To extend the table, David notices how the values of y change in relation to the values of x (for example, as x increases by 1, y increases by 2). He then uses this information to extend the table (see fig. 1.4). However, for David, the variables represent the numbers in the pattern, not the pattern itself. If, as Mr. Lewis believes, David is thinking of the variables x and y as generalizing a pattern, he would write $y = 2x$ to represent the generalized pattern that he has observed.

Suppose, alternatively, that another student, Marcy, views the first table (fig. 1.3) and is interested in the relationship between x and y. She adds a third row to the table and computes the ratio of y to x (see fig. 1.5). After looking at the set of data, Marcy notices a pattern: "If I divide the number in the y-row of the table by the number above it in the x-row, I always get 2." She is not really looking at how the value of y changes as the value of x changes, as David was. She can use variables to express her generalization as $y/x = 2$. For Marcy, the variables represent arbitrary amounts used to generate a pattern.

Because the roles played by variables can depend on the context in which the variables are set as well as how that context is

x	1	2	3	4	5	6
y	2	4	6	8	10	12
y/x	2	2	2	2	2	2

Fig. 1.5. Marcy's extension of David's table

viewed, it can be difficult to arrive at a consensus about the particular role that a variable plays. Regardless of what interpretation is given to a variable, it is important to develop an appreciation for the complexities associated with a thorough understanding of variables.

Understanding what variables represent

Essential Understanding 3b. *A variable represents the measure or amount of an object, not the object itself.*

Consider a problem about a set of objects, such as apples. Depending on the problem, a variable in such a problem might represent many things. It could represent the number of apples, the weight of the apples, or even the cost of the apples. However, the variable cannot represent the apples—the objects themselves. As another example, consider the statement, "There are three feet in one yard." Some might translate this statement incorrectly as "$3f = y$," where f represents the number of feet and y represents the number of yards. In addition to (or maybe because of) making a literal left-to-right translation of the statement, they might consider the variables f and y as representing *feet* and *yards*, respectively, rather than the equivalent *number* of feet and *number* of yards in the length of the object being measured. This idea is complicated by the fact that sometimes a quantity may be abbreviated in a way that confuses a variable with a label. For example, although it is natural to write "3 meters" as "3 m," "m" acts as a label, not a variable. As illustrated in figure 1.6, the *number* of yards must be multiplied by 3 to equal the *number* of feet. Hence, $f = 3y$ where f represents the number of feet and y represents the number of yards is one correct way to represent the relationship, "There are three feet in one yard."

Number of yards (y)	1	2	3	4	5	6	7
Number of feet (f)	3	6	9	12	15	18	21

Fig. 1.6. Comparing the number of feet to the number of yards

Using repeated or different variables

Essential Understanding 3c. The same variable used more than once in the same equation must represent identical values in all instances, but different variables may represent the same value.

Within a given expression or equation, the value represented by a distinct variable must be the same for every occurrence of that variable. For example, $x = 3$ is the only possible solution for the equation $x + x = 6$. Furthermore, when evaluating the expression $x + x + 5$, if we determine that $x = 3$ the first time x occurs, it must also equal 3 in its second occurrence. Similarly, when solving the equation $x + 2 = 2x - 6$, we know that the value represented by the variable x on the left side of the equation must equal the value represented by x on the right side of the equation.

In contrast, when determining all the possible whole numbers x and y such that $x + x + y = 9$, we find that not only is $x = 2$ and $y = 5$ a solution ($2 + 2 + 5 = 9$), but so is $x = 3$ and $y = 3$ ($3 + 3 + 3 = 9$). Similarly, $a = 3$ and $b = 3$ is a possible solution for the equation $a + b = 6$. As these examples illustrate, different variables in the same equation may represent the same value.

The role of a variable in a specific context

Essential Understanding 3d. The same variable may play one or more roles within a given application, problem, or situation.

When examining the equation $5(x + 3) = 20$, we are likely at first to consider that the variable x is playing the role of a fixed but unknown quantity. The goal of solving this equation is to try to find the value of x that makes the equation true. Although we could use various ways to find this value (see the discussion of Essential Understanding 2b on solving equations by using relational thinking strategies), it is common to use the distributive property to transform the equation to the equivalent form $5x + 15 = 20$.

However, when we use the distributive property to write $5(x + 3)$ as $5x + 15$, we change the role played by the variable x. We no longer think of x as a fixed but unknown quantity. Now the variable x serves more as an arbitrary or abstract placeholder in an algebraic process. The goal at this stage is not to determine the value of x. We are simply applying an algebraic process (use of the distributive property) to rewrite the expression $5 \times$ (something $+ 3$) as ($5 \times$ something) $+ (5 \times 3)$.

Essential Understanding 2b Equations can be reasoned about in their entirety rather than as a series of computations to execute.

Variables as representing discrete or continuous quantities

Essential Understanding 3e. *A variable may represent either a discrete or a continuous quantity.*

Regardless of the role that a variable plays in a particular problem, it may represent either a discrete or a continuous quantity. Discrete quantities are reflected by units that cannot (or customarily would not) be broken up. For example, if you have a group of pencils, you can count the pencils one by one. Each pencil thus represents one unit; it would not ordinarily be feasible to break the pencil into smaller pieces. In the Squares and Vertices problem presented later in this chapter (see fig. 1.11 on p. 48), the number of squares, s, and corresponding number of vertices, v, can take on an infinite number of values, but these values will always be whole numbers that cannot be broken up into smaller units. That is, in the situation for this problem, we cannot have a fractional number of squares or vertices.

A variable might also represent a *continuous* quantity. Consider the height of water in a beaker that is 10 centimeters tall. The height of the water can take on any value between 0 centimeters (if the beaker is empty) and 10 centimeters (if it is full). That value might not be a whole number, or even a rational number (although it will be a real number). We cannot say, however, that the height of the water can take on only a certain distinct set of values (for example, 1 centimeter, 2 centimeters, 3 centimeters, and so forth). As another example, consider the height of a child. A child's height can take on any value within an appropriate range of values and does not skip values within that range. That is, a child cannot grow from 30 inches to 40 inches without growing through all of the (infinite) number of values between 30 and 40 inches. Measures such as height (length), area, and volume are said to be *continuous*.

As we have seen, there is no single definition of *variable*. Variables can be interpreted in many different ways: as (1) symbols in a generalized pattern, (2) fixed but unknown numbers, (3) quantities that vary, (4) parameters, and (5) abstract placeholders in an algebraic process. Although students in elementary grades are not expected to recognize these forms in a particular task, teachers need to have a meaningful understanding of variables that allows for flexibility in interpreting them and an awareness of their proper uses. Variables provide an efficient way to communicate algebraic relationships and solve many mathematical problems. In this sense, the use of variables is an indispensable mathematical tool.

Using Quantitative Reasoning to Generalize Relationships: Big Idea 4

Big Idea 4. *Quantitative reasoning extends relationships between and among quantities to describe and generalize relationships among these quantities.*

Quantitative reasoning is an important part of the development of a deep understanding of algebraic ideas. It involves identifying relationships among quantities that have been measured or counted in some way. These relationships can be expressed numerically or with variables. The previous big ideas discussed in this chapter have focused on number and operations, equations, and variables. In this section, we apply those ideas to other situations and extend our thinking to inequalities and to using quantitative reasoning to represent relationships in different ways.

How two quantities relate to each other

Essential Understanding 4a. *Two quantities can relate to each other in one of three ways: (1) they can be equal, (2) one quantity can be larger than the other, or (3) one quantity can be smaller than the other.*

Understanding how two quantities relate to each other is fundamental to quantitative reasoning. To think about the ways in which two quantities can be related, consider the situation in Reflect 1.9.

> ### Reflect 1.9
>
> Temeisha found that the volumes of two glasses were not equal. In what different ways can she describe the relationship between the volumes?

The initial stages of quantitative reasoning begin by identifying the attributes of objects or sets that can be measured or counted. The attributes may be designated as either discrete or continuous. For example, as we described in our discussion of Essential Understanding 3e, a discrete measure might be the number of pencils one has, and a continuous measure might be the volume of liquid in a beaker.

We do not often think of these counts or measures—particularly measures of physical attributes such as volume—as playing a role in the understanding of topics related to algebra. However, measurement topics can provide a context in which students can model relationships in ways that link concrete representations to

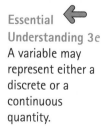

Essential Understanding 3e
A variable may represent either a discrete or a continuous quantity.

more abstract ones while they are developing an understanding of significant mathematical properties. The representations of these relationships provide a strong foundation for algebra not only because of their use of variables, but also because of their depiction of the interactions between and among quantities in generalized forms (Dougherty 2008).

Although attributes of an object may be measured or counted by using a specific unit (for example, inches), a qualitative comparison of measurable attributes may be made without using any units. For example, it is possible to compare the length, area, volume, or mass of two objects without attaching any units to the comparison. Instead, a qualitative comparison can be made by direct comparisons. Qualitative comparisons are powerful because they allow us to explore generalized relationships among quantities, and doing so is an important characteristic of algebraic thinking. For example, when directly comparing length, we can align the two objects so that it is possible to determine which one is longer or shorter without actually measuring their lengths (see fig. 1.7).

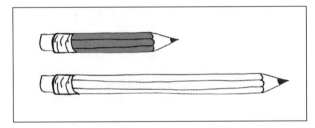

Fig. 1.7. Comparison of lengths of two pencils

As we make these direct comparisons of the lengths, we can observe that the lengths may be equal or unequal. We can describe that relationship in natural language by saying that the two lengths are or are not the same. If they are not the same, we can go further and say that the length of one pencil is less than the length of the other pencil. We can extend the natural language by using variables to represent the lengths of the pencils. If we assign O to represent the length of the orange pencil in figure 1.7 and W to represent the length of the white pencil, we can use these variables efficiently, as described in Big Idea 3, to write $O \neq W$ and $W \neq O$. These representations tell us that the two lengths are not the same (not equal), but these statements do not tell us which length is longer (or shorter). To provide this additional information, we could also write $O < W$ and $W > O$, two equivalent statements.

In fact, generally speaking, whenever we compare two quantities—say, lengths E and K—we will find one of three things: one of the quantities is equal to the other ($E = K$), one is less than other ($E < K$), or one is greater than the other ($E > K$). This statement is known as the *trichotomy property*.

In Reflect 1.9, Temeisha could have written various statements to describe differences in the volumes of the two glasses. For example, if she let H represent the volume of one of the glasses and G represent the volume of the other glass, she could have written $H \neq G$, $G \neq H$, and $G < H$ or $H > G$ (assuming the glass with volume H has more liquid). In the first two statements, it is not possible to tell which of the two volumes is larger (or smaller). The latter two statements are strict inequalities because they indicate which of the measures is greater (or smaller).

It is also possible to use a different representation to show the relationship between volume H and volume G. Diagrams in the form of line segments offer an especially useful means by which the relationship can be shown. Assume that volume H is greater than volume G. Figure 1.8 represents two unequal lengths and shows that volume H is greater than volume G by some amount, which, in this case, is not specified. Line segments can be used to show relationships that involve area, volume, mass, or length.

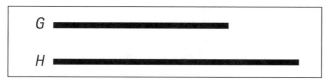

Fig. 1.8. Diagram of the relationship of lengths H and G

What if the two quantities had been equal? Would it be possible to write more than one statement about the relationship? Suppose that we had two groups of marbles, and each group contained the same number of marbles. We could say that D represents the number of marbles in the first group and that R represents the number of marbles in the second group. The statement $D = R$ would represent the relationship between the numbers of marbles in the two groups. It is also possible to write $R = D$ to express the same relationship. This illustrates what is called the *symmetric property of equality*: If $D = R$, then $R = D$.

The symmetric property of equality is important in learning algebra because it allows one to see and express relationships in multiple ways. For example, given the equation $3 \times 7 = $ ___ , one may suppose that the only correct way to write the equation is $3 \times 7 = 21$. However, understanding that this equation can also be expressed as $21 = 3 \times 7$ focuses attention on the relational aspect of the equals sign rather than on the equals sign as an operation symbol (see Essential Understanding 2*a*). The development and use of properties such as the symmetric property through direct qualitative comparisons, including expressions of them in symbolic form, underscores our earlier point that reasoning about quantities can support algebraic understanding.

Essential ⬅
Understanding 2*a*
The equals sign is a symbol that represents a relationship of equivalence.

Reasoning about multiple quantities

Essential Understanding 4b. *Known relationships between two quantities can be used as a basis for describing relationships with other quantities.*

Identifying and describing how two quantities relate form the foundation for more sophisticated thinking about relationships and properties. The trichotomy property is an important property that serves as the springboard for quantitative reasoning, but this property involves only two quantities. What if more quantities are involved? To begin thinking about this question, consider Cameron's comparison of two volumes in Reflect 1.10.

Reflect 1.10

Cameron measured two volumes *A* and *K*. To represent the relationship between the two volumes, he wrote *A* > *K* by *L*.

What might Cameron mean by this statement?

Suppose that we have two containers with volumes J and F, which we know are equal. Then we can write the statement $J = F$ or $F = J$. Suppose that we also know that volume F is equal to volume P of another container. What can we say about the relationship between volumes J and P? Symbolically, we would represent this by the statement, "If $J = F$ and $F = P$, then $J = P$." This is known as the *transitive property of equality*. The means by which we determined the relationship between J and P was not linked to any method of direct measurement. Rather, we used the relationships that we had established between J and F and between F and P to determine the relationship between J and P. We could make that determination on the basis of an indirect comparison that did not involve any physical manipulation of the actual objects.

Construction of indirect comparisons is a bridge between quantitative reasoning and algebraic thinking (Thompson 1988). By making and expressing comparisons between two quantities symbolically, one can act on the quantities without having to manipulate the physical materials (Dougherty 2010). Actions on symbolic expressions of concretely represented quantities prepare children to act on symbolic statements that represent numerical or abstract quantities. For example, if $3 + x = n$, and $n = 3 + 5$, then we know that $3 + x = 3 + 5$ (and can conclude that x must be 5 without doing other algebraic manipulations).

Although we used variables to describe the transitive property of equality, we can also use numbers to write examples of transitive relationships. For example, if we know that $8 \times 3 = 24$ and

24 = 4 × 6 then, by the transitive property of equality, we can conclude that 8 × 3 = 4 × 6. The symbolic statements in the earlier paragraph are a more generalized form of the specific case illustrated here with numbers. What is true about both situations—symbolic and numeric—is that reasoning about quantities where no specific measure or count is used develops a critical flexibility in expressing and applying relationships between these quantities—an important characteristic of algebraic thinking.

The previous examples of Essential Understanding 4b have been with equivalence relationships, but not all relationships involve three quantities that are equal. A relationship among three rectangular areas A, T, and W may be as follows: $A = T$ and $T > W$. What can we say about the relationship between areas A and W? In one sense, area T becomes the unit of comparison. Area T is compared to area A and to area W. Given that area T and area A are equal, we can substitute A for T and thus write $A > W$. With experience in using direct comparisons to create algebraic statements about relationships, we are then more able to reason about the relationship between areas A and W without doing any comparison.

The experience of performing direct comparisons with physical materials and then representing them with algebraic (symbolic) statements builds a conceptual foundation that allows us to act on these symbolic statements. When symbolic or algebraic statements are introduced with no specific context, little meaning can be attached to them. However, when relationships have a context that involves tangible and identifiable attributes, the meaning that is established leads to a stronger understanding when statements have no explicit contexts, such as $2(3 + x) = 12$. In this statement, it would be possible to determine the value of x without doing any formal algebraic manipulation. The thinking could be something similar to the following: "Twice some quantity is 12, so the quantity must be 6, since 2 × 6 is 12. The sum of 3 and some number is 6, so the number must be 3 because 3 + 3 = 6." (See the discussion of Essential Understanding 2b on using relational thinking strategies.)

We can elaborate other relationships that are initially expressed in one or more inequalities. For example, suppose that we have measured two areas, represented the two areas as U and Z, and determined that $U < Z$. We can then indicate the difference—say, area N—that represents the amount by which area U is less than area Z. Thus, we can write the statement $U < Z$ by N (recall Reflect 1.10). This gives rise to other statements that represent this relationship. For example, we can write the following:

$$U + N = Z$$
$$Z - N = U$$
$$N = Z - U$$

We have expressed the relationships among areas U, N, and Z

Essential ⬅
Understanding 2b
Equations can be reasoned about in their entirety rather than as a series of computations to execute.

symbolically and through an actual measure or comparison of the areas. In addition, we can show the relationships with line segments, using lengths of the segments even though we initially worked with areas. Line segments afford us the opportunity to depict a relationship without establishing numerical values for the quantities. They are a step removed from the physical materials and offer a way to record a relationship between or among quantities without using symbols.

Thus, the relationship between areas U and Z is shown in figure 1.9 with line segments. From this representation, it is clear that area Z differs from area U by some amount. In fact, the amount by which they differ is represented by the length of line segment Z that extends beyond line segment U. Diagrammatically, this illustrates that this difference quantity could be either added to the line segment representing area U or subtracted from the line segment representing area Z to create two equal quantities.

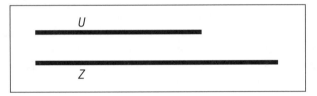

Fig. 1.9. A line segment diagram showing the relationships of areas Z and U

As indicated in the equations that relate U and Z, we designated the difference between U and Z to be area N. In figure 1.10, this difference, area N, can be represented by the length of the line segment that, when added to the length of the shorter segment, creates two equal lengths.

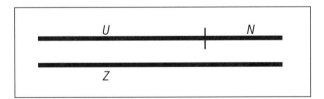

Fig. 1.10. A line segment diagram showing the relationships of areas Z and U and the difference, area N

This example represents the way in which two quantities that are unequal can be made equal. In general, it illustrates that the amount by which one quantity is larger than the other can be added to the lesser quantity or taken away from the larger quantity to create a new quantity that is now equal to the other.

This flexibility in working with these relationships extends to other contexts. For example, if x is greater than 4 by 7, the following equations can be written:

$$4 + 7 = x$$
$$x - 4 = 7$$
$$x - 7 = 4$$
$$x = 7 + 4$$

Placing this in a strictly numerical context, we can see that it is possible to write the statement, "137 is greater than 122 by 15." This statement, modeled after the more general form above, indicates that the difference between 137 and 122 is 15. Similarly, we can write statements that illustrate this relationship:

$$137 - 15 = 122$$
$$122 + 15 = 137$$
$$137 - 122 = 15$$
$$137 = 122 + 15$$

Consider other possibilities. If we start with two volumes that are equal, such as volume Y and volume S, we can write the statement $Y = S$. We can change both volumes by the same amount and still maintain an equality relationship. For example, if volume C is added to both Y and S, the equality relationship continues to hold: $Y + C = S + C$. Similarly, we can remove the same volume, say volume P, and also maintain the equality, $Y - P = S - P$.

This flexibility in representing equality relationships sets the stage for solving equations. Instead of simply relying on a rule, we see that equality is maintained when the same quantity is added to (or subtracted from) the quantities that are being compared. Thus, when we are faced with an equation like $27 = x + 14$, one method that we might use to solve the equation would involve subtracting (removing) 14 from both quantities, 27 and $x + 14$, so that we would maintain the equality relationship. Or, alternatively, knowing that it is possible to rewrite the equation, we could also write the following equivalent equations:

$$27 = 14 + x$$
$$27 - 14 = x$$
$$27 - x = 14$$

Then, instead of being forced to subtract 14 from both sides of the equation $27 = 14 + x$, we could use the equivalent equation, $27 - 14 = x$, to solve it.

These ideas about the relationships of quantities are important steps in developing the ability to make sense of problems that have no story or real-world context. Given an equation such as $3x + 4 = 8 - x$, problem solvers then understand that the relationship expressed indicates that the two quantities or amounts, such as $3x + 4$ and $8 - x$, are equivalent. Without this relational understanding of the equals sign, they place a limited interpretation on the solution to this equation as the value obtained by solving the equation and not the value that makes both quantities the same. The distinction may appear small, but it is critical to a conceptual approach to understanding the solution and the process of solving equations.

The approach of quantitative reasoning taken here begins with generalizations about relationships between and among quantities. Using this perspective, quantitative reasoning can be seen as the vehicle by which to develop the ability to see quantity as something that can vary.

Functional Thinking as a Path into Algebra: Big Idea 5

Big Idea 5. *Functional thinking includes generalizing relationships between covarying quantities, expressing those relationships in words, symbols, tables, or graphs, and reasoning with these various representations to analyze function behavior.*

Before we discuss essential understandings associated with Big Idea 5, we would like to explain briefly why we regard functional thinking as a big idea of algebra in grades 3–5. The study of functions is a vast area of mathematics that does not need to be cast in a supporting role in relation to algebra. However, it can provide an important entry point into algebra (Carraher and Schliemann 2007). In particular, as a context for developing algebraic understanding, it allows us to cast algebra as the use of different representations to make sense of relationships in (quantitative) situations (see Ellis [in press] and Kieran [1996]). Interpreting algebra in this way, we focus our attention here on those mathematical aspects of functions that can also develop algebraic understanding: (1) generalizing relationships, (2) expressing those relationships in multiple ways, and (3) reasoning with those generalizations. Consider a brief illustration.

We can extend the arithmetic task of finding the number of legs on 20 chairs, where each chair is assumed to have 4 legs, to an algebraic task of finding a generalized relationship or rule for the number of legs on *any* number of chairs (see the Chair and Leg problem in Reflect 1.2). Moreover, describing this relationship can help build an understanding of symbolic, algebraic notation. Although we might first describe the relationship in natural language ("The number of legs is four times the number of chairs"), we can also express it symbolically as $L = 4C$, where C represents the number of chairs and L represents the number of legs. Generally, the choice of letters is arbitrary, although pragmatic reasons might guide the selection of a particular letter. Also, although mathematics has some conventions regarding the expression of variables in uppercase or lowercase, the choice of uppercase here is arbitrary as well.

Natural language and algebraic notation are just two ways in which we might represent a functional relationship. We might also construct a table or a graph to represent the relationship. As we will discuss in connection with Big Idea 5 later, different representations offer different perspectives on how we might interpret a function and reason about it, and it is important to understand these different representations and how they are connected.

Finally, if we consider chairs with 5 legs, we might then reason from the generalized relationship $L = 4C$ to determine the new relationship to be $L = 5C$. That is, the only change in the problem statement (the assumption that chairs have 5 legs instead of 4) is reflected by a change in the relationship from $L = 4C$ to $L = 5C$. In this sense, functions help us learn to *reason about* structure, not just notice it.

These ideas suggest how the study of functions in elementary grades can leverage some of the core aspects—or essential understandings—of early algebraic thinking. We open our discussion of these essential understandings with a look at the special nature of functions.

A function as a relationship between two sets

Essential Understanding 5a. *A function is a special mathematical relationship between two sets, where each element from one set, called the* domain, *is related uniquely to an element of the second set, called the* co-domain.

Consider the following task (adapted from Anku [1997]):

The Squares and Vertices Problem

Suppose that you were given the configurations of squares shown in figure 1.11. How would you describe the relationship between the number of squares and the number of vertices in each configuration? How would you describe this relationship for configuration 100—that is, for a configuration with 100 squares? How would you describe this relationship for configuration n?

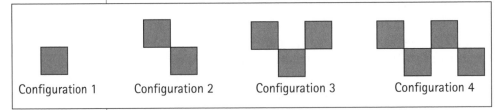

Configuration 1   Configuration 2   Configuration 3   Configuration 4

Fig. 1.11. The Squares and Vertices problem

One way to organize the information given in this task is by recording in a function table the number of squares and the corresponding number of vertices for each configuration. For example, in the table in figure 1.12, we let s represent the number of squares and v, the number of vertices, and we record the number of squares and corresponding number of vertices for configurations 1–4. Reflect 1.11 examines the variables s and v.

stop reflect

See Reflect 1.11 on p. 49.

s	v
1	4
2	7
3	10
4	13

Fig. 1.12. A function ta ares and Vertices problem

Reflect 1.11

What role do the variables play in the Squares and Vertices problem?

The data in this table depict a special kind of relationship that we describe as a *function*. A function is a relationship in which each element in one set is paired uniquely with a single element in a second set. In the Squares and Vertices problem, each element in the set $\{1, 2, 3, 4, ...\}$, called the *domain* of the function, is paired with a single element in the second set $\{4, 7, 10, 13, ...\}$, called the *co-domain* of the function. In other words, for each number of squares, there is a *unique* number of vertices. For example, configuration 2 has exactly 7 corresponding vertices. It cannot have, say, both 7 and 14 corresponding vertices. In general, any particular configuration of squares will always have exactly *one* corresponding number of vertices. This unique correspondence is what distinguishes a function from other relationships between elements in two sets.

It is important to note that in addition to *domain* and *co-domain* of a function, we can talk about the *range* of a function. The range is the subset of the co-domain that includes only those elements of the co-domain that are "used." For example, the quadratic function given by $y = x^2$ typically has the real numbers as its domain and can be said to have the real numbers as its co-domain, but its range is the nonnegative real numbers. With more complex functions than those used in elementary grades, it is sometimes difficult to identify the range. In this book, we choose to work with the more inclusive notion of co-domain rather than range.

The elements in the domain and co-domain of a function are often (but not necessarily) numbers, and the resulting relationship might not be an intuitive pairing between elements. The relationship might not even be one that can be described as a sequence of mathematical operations (for example, "the number of chair legs is four times the number of chairs"). Consider the following two sets: $\{\triangle, \Diamond, \square\}$ and $\{A, B, C, D\}$. We can define a functional relationship between these sets by setting up the following correspondence: $\{(A, \square), (B, \Diamond), (C, \Diamond), (D, \triangle)\}$. Here, A is paired with $\square$, B is

Recall that different roles or meanings associated with *variable* are the focus of Essential Understanding 3a.

← Essential Understanding 3a
The meaning of variable *can be interpreted in many ways.*

Although it is true for the function under discussion that the *n*th element of the first set is paired with the *n*th element of the second set, that pairing should not be assumed for every function.

For a more detailed discussion of function and relationships that are not functions, see *Developing Essential Understanding of Functions for Teaching Mathematics in Grades 9–12* (Cooney, Beckmann, and Lloyd. 2010).

paired with ◇, C is paired with ◇, and D is paired with Δ. What makes these pairings form a function is that each element of the set {A, B, C, D} is paired with exactly one element from the set {Δ, ◇, □}. For example, A is not paired with both □ and Δ. In this example, the domain and co-domain are not sets of numbers. Moreover, the relationship formed between the sets is not intuitive, nor can it be described by a sequence of mathematical operations. That is, we do not "do" something mathematically to A to produce □.

In early algebra, however, the functions studied are typically based on relationships between sets of numbers, they are well behaved, and they have an underlying closed-form mathematical rule (for example, $L = 4C$). This is not necessarily the case in a more formal study of functions in later grades, when not all functions can be expressed in this way. Finally, though algebra instruction in the elementary grades typically does not explicitly define concepts such as function or even domain or co-domain of a function, it is important for elementary teachers to appreciate subtle distinctions in these concepts and the special nature of a functional relationship.

A covariational perspective

Essential Understanding 5b. *Functions can be viewed as tools for expressing covariation between two quantities.*

Although the characterization of functions in Essential Understanding 5a reflects how functions have historically been treated in school mathematics, it does not fully convey the co-variation that underlies many functional relationships (see Ellis [in press]). For example, the functional relationship underlying the Squares and Vertices problem can technically be thought of as a relationship where each element from the set containing the number of squares in each configuration (that is, {1, 2, 3, 4, 5, ...}) is related uniquely to an element in the set containing the number of vertices for each configuration (that is, {4, 7, 10, 13, 16, ...}).

But we might also think of the relationship in terms of how the quantities vary in relation to each other: As the number of squares increases by 1, the number of vertices increases by 3. The importance of a covariational perspective on functions is that it focuses on functions as tools for representing covariation in quantities, an idea that is an important precursor to the idea of rate of change and other fundamental concepts of calculus (Smith and Confrey 1994). Typically, function tasks used in early algebra, such as the Squares and Vertices problem, can be viewed from a covariational perspective.

A detailed discussion of different types of functions appears in *Developing Essential Understanding of Functions for Teaching Mathematics in Grades 9–12* (Cooney, Beckmann, and Lloyd 2010).

 Essential Understanding 5a
A function is a special mathematical relationship between two sets, where each element from one set, called the domain, *is related uniquely to an element of the second set, called the* co-domain.

The nature of variables in a function

Essential Understanding 5c. *In a functional relationship between two covarying quantities, a variable is said to be either* independent *or* dependent *and will represent either a discrete or a continuous quantity.*

As we will explore in more detail later, the functional relationship, or function rule, in the Squares and Vertices problem can be expressed as $v = 3s + 1$, where v is the number of vertices and s is the number of squares. Because v is expressed in terms of s, the number of vertices, v, is said to depend on the number of squares, s. In this case, s is characterized as the *independent variable* and v as the *dependent variable*. This is a reasonable choice, since the problem itself suggests that the number of vertices will change as the number of squares in each configuration changes. However, the choice of independent and dependent variables is somewhat arbitrary. For example, we could re-express $v = 3s + 1$ as

$$s = \frac{v-1}{3}$$

to represent s as a function of v. Typically, however, function tasks in early algebra afford natural choices for the independent and dependent variables.

Moreover, variables in function tasks used in early algebra typically involve discrete quantities because these best reflect young children's experiences with number. The values of these variables might arise from concrete situations involving the number of legs on a spider, the number of cuts of a piece of string, the number of people attending an event, and so forth. However, it is important for elementary teachers to understand that this is not the case for all functions. As the study of functions deepens beyond elementary grades, variables often represent continuous quantities, such as the height of a child or the volume of liquid in a container, and are sometimes based on "messy" real-world data (Common Core State Standards Initiative 2010; NCTM 2000).

Different types of relationships in a function

Essential Understanding 5d. *In working with functions, several important types of patterns or relationships might be observed among quantities that vary in relation to each other: recursive patterns, covariational relationships, and correspondence rules.*

As students' work with pairs of quantities that vary in relation to each other expands and their mathematical experience grows, they will begin to observe types of patterns or relationships in the way in which the quantities vary with respect to each other. Three types that are very important and useful are recursive patterns, covariational relationships, and correspondence rules (Smith 2003).

Recursive patterns describe variation in a single sequence of values. For example, in the Squares and Vertices problem, one might identify a pattern in how the number of vertices changes within the ordered sequence 4, 7, 10, 13, ... The pattern might be expressed informally in a statement such as, "Start at 4 and add 3," or, more simply, "Add 3," with the starting value, 4, understood (see fig. 1.13).

s	v
1	4
2	7
3	10
4	13

$\Big)$ +3

Fig. 1.13. A recursive pattern for the Squares and Vertices problem

Generally, a recursive pattern indicates how to obtain a number in a sequence from the previous number or numbers. However, because recursive patterns attend to change in only one variable, they have limited applicability. In particular, gathering information about the function value for a large value of the independent variable requires calculating all corresponding prior function values. For example, finding the number of vertices for a configuration of 100 squares (configuration 100, or the case when $s = 100$) would require knowing the number of vertices for each of the preceding 99 configurations of squares. Calculating this by hand would be inefficient. By using recursive patterns, more sophisticated tools such as computer software allow function values to be calculated easily, but these tools are not typically used in early algebra.

Another limitation of recursive patterning is that it does not involve any analysis of the independent variable: Identifying a pattern in the ordered sequence 4, 7, 10, 13, ... as "start at 4 and add 3" does not require any information about the number of squares or how a particular number of squares relates to the corresponding value in the sequence denoting the number of vertices. That is, it does not involve attending to *covariation* between quantities. Thus, while finding recursive patterns is often a first step in making sense of data, functional thinking requires looking beyond a single sequence of values and understanding how quantities vary *in relation to each other*, or *covary*. As we describe next, this might occur in two ways that are particularly relevant for early algebra.

Covariational thinking involves analyzing how two quantities vary in relation to each other and keeping that variation explicit in the description of the function. With the Squares and Vertices problem, covariational thinking might be exhibited in the following description of the relationship: "As the number of squares increases by 1, the number of vertices increases by 3" (see fig. 1.14).

s	v
1	4
2	7
3	10
4	13

+1 ↻   ↻ +3

Fig. 1.14. Covariation in the Squares and Vertices problem

Alternatively, the relationship between the number of squares and the number of vertices might be described in the statement, "As the number of vertices increases by 3, the number of squares increases by 1." One child, Hector, described this covarying relationship in the following way: "Every time you get one more number, you get three more than that each time." Although Hector's language still has some ambiguity, it clearly shows that he is attempting to express how the two quantities vary in relation to each other.

Understanding coordinated change in quantities—covariation—is an important aspect of functional thinking. However, a complete understanding of functions should include a broader exploration of the relationships between two quantities than covariation alone provides. A *correspondence relationship* is a correlation between two quantities expressed as a function rule. In identifying a correspondence relationship in the Squares and Vertices problem, one student first described how he found the number of vertices: "I multiplied the number of squares times three and added one to it." Building on this student's idea, another student, wrote the correspondence relationship symbolically as $s \times 3 + 1 = v$, where s represented the number of squares and v the number of vertices. As this suggests, a correspondence relationship entails identifying a generalized relationship between two quantities. In other words, it goes beyond describing variation in a single sequence of values (such as "add 3," as in a recursive pattern) or even describing how one quantity varies in relation to another (for example, "As the number of squares increases by 1, the number of vertices increases by 3").

Rules for correspondence relationships are particularly useful because they allow us to determine information about a specific function value easily without knowing other function values. For example, we can use the rule $v = 3s + 1$ to quickly calculate the number of vertices for a configuration of 1000 squares to be $(3 \times 1000) + 1 = 3001$, without any information about other function values. If we only had a recursive pattern and we wanted to calculate the number of vertices for a configuration of 1000 squares, we would need to know the number of vertices for each configuration of squares up to 999 squares.

For many functions, we can identify a rule by looking for relationships in a function table or by reasoning from the context of the problem. Although sophisticated techniques are often needed in more advanced mathematics, function data used in early algebra can often be analyzed by using these methods. Identifying a rule from the function table often involves guess-and-check methods. However, we can be more systematic if we reason from the context of the problem. Problem contexts that involve geometric patterns can be very important in visualizing changes in a relationship, and visual strategies may be more powerful than numerical strategies alone for identifying functional relationships (Rivera and Becker 2005).

In the Squares and Vertices problem, for instance, we can find a rule by looking carefully at how the count for the number of vertices varies as the number of squares changes in subsequent configurations. For example, each time a square is added to an existing configuration of squares, three additional vertices are added in the new configuration (see fig. 1.15).

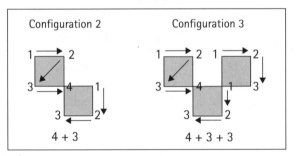

Fig. 1.15. Systematically counting the number of vertices in configurations 2 and 3

If we track this pattern for each of the configurations, we can express the number of vertices in configurations 1 through 4 as 4, 4 + 3, 4 + 3 + 3, and 4 + 3 + 3 + 3. The constant in each sum is 4, while the number of times 3 appears varies according to the configuration number (or number of squares). In particular, the number of 3's in the sum of vertices for a particular configuration is one less than the number of squares, or $s - 1$. Thus, the sum of 3's for a figure with s squares can be expressed as $3 \times (s - 1)$. Combining this information, we can describe the number of vertices as $v = 4 + (3 \times (s - 1))$. Using the distributive property, we can simplify this to $v = 4 + 3(s) - 3(1)$, or $v = 3s + 1$. Furthermore, we can interpret this rule in terms of the problem context by noting that, for each configuration, there is always one vertex from the original square and three vertices *for each number of squares* in the configuration (that is, $3s$). The sum, $3s + 1$, then gives us the total number of vertices.

Function rules are an important tool for reasoning about function behavior. In more advanced mathematics, rules representing functions can be analyzed by using tools of calculus to determine rates at which quantities change or accumulation over time. Real-world phenomena such as changes in weather patterns, the stability of physical structures when subjected to natural forces, and the growth of a virus within a population all call for the use of functions to model their behavior and, subsequently, an analysis of these functions to understand changes in the behavior. A more advanced study of functions extends into these areas. Thus, there are significant mathematical reasons why early algebra should begin to look beyond identifying simple recursive patterns and to study covariational and correspondence relationships. Reflect 1.12 wraps up this discussion with a practical application.

Reflect 1.12

Using the Squares and Vertices problem as a guide, design a function task that you might use with your students. Identify a recursive pattern, covariational relationship, and a correspondence rule in your function.

Different representations of functions

Essential Understanding 5e. *Functions can be represented in a variety of forms, including words, symbols, tables, and graphs.*

A rich understanding of functions is characterized by *representational fluency*—that is, an ability to represent functions in multiple ways and to navigate flexibly among these various representations. The Representation Standard articulated in *Principles and Standards for School Mathematics* calls for instruction that enables all students to—

- create and use representations to organize, record, and communicate mathematical ideas;

- select, apply, and translate among mathematical representations to solve problems;

- use representations to model and interpret physical, social, and mathematical phenomena. (NCTM 2000, p. 67)

The representations used to identify and reason with functions should reflect this Standard. With the Squares and Vertices problem, we used a function table to organize and record information about numbers of squares and vertices (see fig. 1.12). This tabular representation helped organize the data so that any relationships in them might be more apparent. We then used this table to identify a

recursive pattern in words ("start at 4 and add 3") and a correspondence rule, or function, in words ("the number of vertices
is 3 times the number of squares, plus 1"). We also expressed
our understanding of the function by writing the symbolic rule
$v = 3s + 1$. Furthermore, we could use any of these representations
of the function (words, symbolic rule, or table) to construct its graph
on a Cartesian coordinate system (see fig. 1.16).

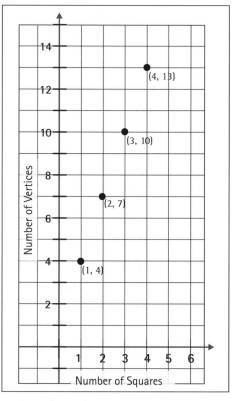

Fig. 1.16. A portion of a graph representing the Squares and Vertices
problem

In the elementary grades, coordinate graphs are typically constructed by plotting points. Some caution should be taken with a
coordinate graph. First, note that only the points are drawn in figure
1.16 (that is, no line is drawn to connect the points). A line could be
interpreted to mean that any point on the line would satisfy the
conditions of this problem, which is clearly not the case.
Technically, all points on the line satisfy the equation $v = 3s + 1$.
For example, the point $(1/3, 2)$ is a point on the line because substituting these values into the equation $2 = (3 \times 1/3) + 1$ gives a true
statement.

However, only whole number solutions are suitable for the
Squares and Vertices problem. For example, you cannot have a
fractional or negative number of squares or vertices. In other

situations, things might well be different—a possibility that Reflect 1.13 asks you to consider. Furthermore, although the graph shows only four points, it can be continued indefinitely to show more and more points, since the number of squares (and, therefore, the number of vertices) can always be increased. In this sense, it is important to understand that the information depicted in a graph does not always give the complete picture. Similarly, a table like that in figure 1.14 does not always tell the whole story.

Reflect 1.13

Can you think of a problem situation in which it would be appropriate to connect the points in a graph?

Representational fluency also involves the ability to "select, apply, and translate among mathematical representations to solve problems" and to use these representations to model and interpret mathematical situations (NCTM 2000, p. 67). Those who have this fluency recognize that one particular representation is not necessarily "better" than another, but that distinct representations provide different ways of seeing a relationship that can facilitate problem solving. To understand this further, first solve the task presented in figure 1.17, and then consider Reflect 1.14, which probes your reasoning strategies and use of representation.

Raymond has saved some money. As a reward, his grandmother offers him two deals:

Deal 1: She will double his money.

Deal 2: She will triple his money and then take away 7 dollars.

Raymond wants to choose the best deal. What should he do? How would you help him determine the best approach? Is one deal always better? Why?

(For simplicity, assume that Raymond's money is only in dollar amounts.)

Fig. 1.17. The Best Deal problem. (Adapted from Brizuela and Earnest [2008].)

Reflect 1.14

How did you represent the relationships given by deal 1 and deal 2? In solving this task, how did you navigate among representations (for example, did you use tables to construct graphs, did you use graphs to construct symbolic rules, or did you do something else)?

Suppose that we let c represent the amount of money that Raymond has. Then, for deal 1, we can represent the amount of money that Raymond's grandmother will give him, or twice the amount c, as $2c$. For deal 2, we can represent the amount of money that she will give him, or triple the amount c, less $7, as $3c - 7$. We can then make both tables and graphs as representations of the two functions, as in figure 1.18.

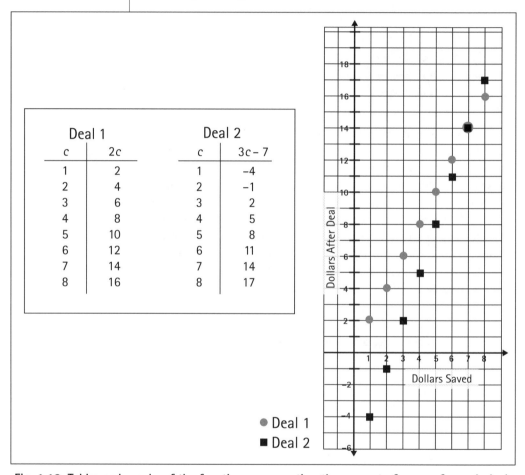

Deal 1			Deal 2	
c	$2c$		c	$3c - 7$
1	2		1	−4
2	4		2	−1
3	6		3	2
4	8		4	5
5	10		5	8
6	12		6	11
7	14		7	14
8	16		8	17

● Deal 1
■ Deal 2

Fig. 1.18. Tables and graphs of the functions representing the amount of money for each deal

Although we have now represented the mathematical problem in multiple ways, the question is, How do we interpret these representations to identify the best deal? Understanding how to interpret different representations of a function is a characteristic of a rich understanding of functions. Reflect 1.15 encourages you to explore possibilities for interpreting our representations of the deals that Raymond's grandmother is offering.

In the tables in figure 1.18, we can see from the function values (or the amount of money that Raymond's grandmother will pay him) that deal 1 is better than deal 2 if the amount of money that

See Reflect 1.15
on p. 59.

Raymond has is less than $7. For each dollar less than $7, the cor-responding function value is greater in deal 1 than its counterpart in deal 2. We can infer this same information from the graphs by observing that if Raymond has less than $7, then each point on the graph for deal 1 is above the corresponding point on the graph for deal 2. Using similar reasoning, we can infer from either the tables or graphs that, if Raymond has 8 or more dollars, deal 2 is better. However, the tables also show that if Raymond has 7 dollars, then both deals provide the same return. This information is conveyed graphically as a point of intersection between the graphs of deal 1 and deal 2.

In summary, these representations illustrate that the deal that Raymond should choose depends on the amount of money that he currently has: If he has less than $7, he should choose deal 1; if he has more than $7, he should choose deal 2; and if he has $7, it does not matter which deal he chooses. As this problem suggests, the advantage of representing a function in multiple ways is that it accommodates different methods of thinking about a mathematical situation, and this flexibility can strengthen understanding of how to interpret function behavior.

Analyzing change to understand function behavior

Essential Understanding 5f. Different types of functions behave in fundamentally different ways, and analyzing change, or variation, in function behavior is one way to capture this difference.

Although functions come in many shapes and sizes, most functional thinking tasks in early algebra depict relationships that are linear, quadratic, or exponential. A detailed treatment of these functions is beyond the scope of this book, where our purpose is to briefly consider distinctions among these types of functions, depending on how they change. An important distinction among them is whether and how they grow (or decay). The study of change in a function's behavior arguably is more firmly rooted in calculus than in algebra, and change and variation are fundamental concepts that are

For a more detailed discussion of linear, quadratic and expo-nential functions, see *Developing Essential Understanding of Expressions, Equations, and Functions for Teaching Mathematics in Grades 6–8* (Lloyd, Herbel-Eisenmann, and Star forthcoming).

Developing Essential Understanding of Functions for Teaching Mathematics in Grades 9–12 (Cooney, Beckmann, and Lloyd 2010) offers an extended discussion of linear, quadratic, exponen-tial, and other types of functions.

formalized in the study of differential calculus. However, analyzing change also gives us opportunities for reasoning with generalizations depicted through symbolic rules, tables, graphs, and so forth, and these experiences support the development of algebraic understanding.

A central characteristic of a *linear* function is that it grows or decays at a constant rate. For example, in the Squares and Vertices problem, the function table shows that each unit increase in the number of squares yields a constant change—in this case, an increase—of 3 vertices. We saw this earlier in the recursive pattern characterized as "start at 4 and add 3" (see fig. 1.13). Said another way, the difference between any two successive function values is always the constant value 3. Because the growth rate is constant, the relationship between the two quantities is said to be linear. *Constant functions* are a special case of linear functions where a unit change in the independent variable yields a change of 0 in the dependent variable. We might interpret this "growth" rate as no growth. The graph of a constant function is a horizontal line.

Functions for which there is not a constant difference between successive function values are said to be *nonlinear*. Two special types of nonlinear functions that are used in early algebra are quadratic and exponential functions. We can understand more about the differences in the behavior of linear, quadratic, and exponential functions by considering the following task:

The Caterpillars' Diet Problem

Three caterpillars are each fed a different diet. Caterpillar A is fed diet A, caterpillar B is fed diet B, and caterpillar C is fed diet C. While the caterpillars are on this diet, each one's length is measured every day for one week and recorded in a separate table, where D represents the day on which the measurement was taken, and L represents the corresponding length, in millimeters, of the caterpillar. Figure 1.19 shows the tables for the three caterpillars. Assuming that no significant factors besides diet affect their growth, which diet—A, B, or C—do you think is the most effective in increasing the length of a caterpillar?

Diet A		Diet B		Diet C	
D	L	D	L	D	L
1	3	1	1	1	2
2	5	2	4	2	4
3	7	3	9	3	8
4	9	4	16	4	16
5	11	5	25	5	32
6	13	6	36	6	64
7	15	7	49	7	128

Fig. 1.19. Tables for the Caterpillars' Diet problem

Reflect 1.16 investigates possible reasoning strategies and representations for solving this problem.

> ## Reflect 1.16
>
> For each of the functions represented in the Caterpillars' Diet problem, identify a recursive pattern and correspondence rule.
>
> Construct the graph for each set of data on the same coordinate axes.
>
> What differences do you notice for each of the diets? Which diet would you choose if you wanted to get the greatest caterpillar growth? Why?

As the function tables suggest, there are clear differences in how much each of the caterpillars grows, depending on its diet (assuming, as the problem asks, that except for diet, no other significant factors affect their growth). As we discussed in connection with Essential Understanding 5d, recursive patterns can help illustrate such differences in growth. For diet A, we can obtain each successive function value (length of the caterpillar) by adding 2 to the previous function value (see fig. 1.20). Because the difference between any two successive function values is a constant value, the function is said to be linear. Later, when we compare graphs of the caterpillars' lengths over time (see fig. 1.22), we will see that the graph of this function is a line.

Essential Understanding 5d
In working with functions, several important types of patterns or relationships might be observed among quantities that vary in relation to each other: recursive patterns, covariational relationships, and correspondence rules.

Diet A	
D	*L*
1	3
2	5
3	7
4	9
5	11
6	13
7	15

+2 (between 3 and 5), +2 (between 5 and 7)

Fig. 1.20. Recursive pattern for diet A

In contrast, the functions that represent diet B and diet C are not linear. For example, if we consider successive differences in function values for diet B, we get (in order of occurrence) *increasing* differences of 3, 5, 7, 9, 11, and 13. For diet C, we get increasing differences of 2, 4, 8, 16, 32, and 64 (see fig. 1.21). We can clearly see that the differences are not a constant value. Moreover, figure 1.19 shows that the length of the caterpillar on diet C increases by the greatest amount over the week in which measurements are recorded. Although the caterpillar on diet B starts out with greater increases in growth, after day 4 this trend reverses, and the caterpillar on diet C begins to show greater increases in growth. With these growth patterns continuing, over time the caterpillar on diet C will grow faster than the other two caterpillars. The caterpillar on diet A will grow the most slowly.

Diet B			Diet C		
D	L		D	L	
1	1	⎫ +3	1	2	⎫ +2
2	4	⎬ +5	2	4	⎬ +4
3	9	⎭	3	8	⎭
4	16		4	16	
5	25		5	32	
6	36		6	64	
7	49		7	128	

Fig. 1.21. Recursive patterns for diet B and diet C

We can see the differences in the caterpillars' rates of growth visually by comparing graphs of the three functions (see fig. 1.22). In particular, the graphs show that after day 4, the points associated with diet C lie above those for the other two diets. After day 2, points associated with diet A lie below points for the other two diets. Ultimately, the choice of which diet to select depends also on how long a caterpillar might eat the particular diet. If the diet can be given for only a few days, then diet B seems to be the best choice. But if the diet can continue for more than 4 days, then diet C appears to yield the best growth.

Note that the points in the graph of figure 1.22 are connected. Recall that a quantity represented by a variable can be either discrete or continuous (Essential Understanding 3e). Because the variable L represents the length of a caterpillar and length is a continuous quantity, it is appropriate to connect the points on the graph.

➡ Essential Understanding 3e
A variable may represent either a discrete or a continuous quantity.

➡ Essential Understanding 5d
In working with functions, several important types of patterns or relationships might be observed among quantities that vary in relation to each other: recursive patterns, covariational relationships, and correspondence rules.

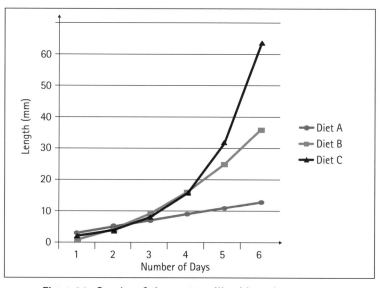

Fig. 1.22. Graphs of three caterpillars' lengths over time

Finally, we can represent these data through a correspondence rule, as suggested in Essential Understanding 5d. Data for diet A can be described by the *linear* relationship $L = 2D + 1$, where L is the length of the caterpillar and D is the number of days on the diet. We can see how to obtain the rules for the nonlinear functions depicted in diets B and C by looking more closely at the lengths and comparing each of these to the number of the day on which it occurs (see fig. 1.23).

Diet B				Diet C		
D	L	L		D	L	L
1	1	1^2		1	2	2^1
2	4	2^2		2	4	2^2
3	9	3^2		3	8	2^3
4	16	4^2		4	16	2^4
5	25	5^2		5	32	2^5
6	36	6^2		6	64	2^6
7	49	7^2		7	128	2^7

Fig. 1.23. Extended function tables for diets B and C

For diet B, the table shows that a particular length can be represented by the square of the number of the day on which the length was taken. For example, on day 3, the length of the caterpillar on diet B is 9 millimeters. We can describe this rule symbolically as the quadratic function $L = D \times D$, or $L = D^2$, where L is the length of the caterpillar and D is the number of days on the diet. For diet C, we can represent a particular length as 2 multiplied by itself D times. For example, the length for day 3 can be obtained by the product $2 \times 2 \times 2$, or 2^3. We can express the rule symbolically as the exponential function $L = 2^D$, where L is the length of the caterpillar and D is the number of days on the diet.

Learning to express functional relationships in symbolic form not only strengthens understanding and facility in the use of a symbolic language—a skill that is so essential to algebra—but as the study of functions deepens, flexibility with symbolic rules also supports analysis of changes in the behavior of complex functions through more sophisticated techniques.

In this section, we have tried to describe how functional thinking can help develop algebraic understanding. In particular, the study of functions involves generalizing relationships in data by shifting attention from relations among particular numbers to relations among sets of numbers. It includes learning to express these generalizations through a variety of representations, including words, symbols, graphs, and tables. It includes reasoning about the inherent structure in these representations and using this structure to solve meaningful problems. It provides a context for understanding the unique role of a variable as a varying quantity. Using all of these aspects of functional thinking is a central part of what it means to "do" algebra.

Finally, it is important to remember that the ideas in this section represent essential understandings that elementary *teachers* need so that they can develop students' algebraic thinking in grades 3–5. Although we do not expect students in these grades to learn formal definitions of terms such as *function* or *independent variable* and *dependent variable*, to identify variables as discrete or continuous, to describe a specific type of relationship as, say, recursive, or to formally distinguish between different types of linear or nonlinear functions, we do know that they can identify recursive patterns, describe covariation, and symbolize correspondence (function) rules. We know that they can represent relationships between covarying quantities in various ways and reason with these representations to solve problems. Our goal in this section has been to identify and discuss knowledge and skills, in addition to these, that would help elementary teachers understand functions in a broader way so that they can appropriately guide students' thinking.

Conclusion

In this chapter, we have identified five big ideas that encompass core algebraic understanding for elementary teachers, beginning with Big Idea 1, which emphasizes the connections between arithmetic and algebra. As we stated in the introduction, the heart of early algebra is in generalizing mathematical ideas, representing and justifying generalizations in multiple ways, and reasoning with generalizations (Kaput 2008). Although we discussed each of the big ideas separately, it is by understanding them in a connected way that teachers are prepared to engage elementary school children fully in early algebra. A few examples of the important connections among the big ideas follow.

Developing a relational understanding of the equals sign, coupled with an understanding of variable, is fundamental to understanding algebra. A relational understanding of the equals sign enables us to represent many important mathematical ideas with equations (Big Idea 2). Understanding variables (Big Idea 3) allows us to represent these ideas in a general sense without being confined to specific numerical examples.

Typically, we think of the importance of using equations to represent ideas within the context of arithmetic, but these representations are also important to reasoning about quantities (Big Idea 4). Exploring and representing comparisons of continuous quantities such as volume or length can strengthen the development of a relational understanding of the equals sign.

Big Idea 1

Addition, subtraction, multiplication, and division operate under the same properties in algebra as they do in arithmetic.

Big Idea 2

A mathematical statement that uses an equals sign to show that two quantities are equivalent is called an equation.

Big Idea 3

Variables are versatile tools that are used to describe mathematical ideas in succinct ways.

For example, comparisons of quantities that are not counted or measured can help develop an understanding of important properties of equality (such as the symmetric property of equality), and this understanding can strengthen a relational understanding of the equals sign. Suppose that we compare two unspecified volumes, Y and S, and find that the volumes are the same. As we discussed in connection with Big Idea 4 on quantitative reasoning, we might express this relationship as $Y = S$. Moreover, when we explore different ways to express the relationship, we come to understand that $Y = S$ is equivalent to $S = Y$. In turn, if we understand that if $Y = S$, then $S = Y$ (the symmetric property), we are also likely to understand that a numerical statement such as $4 \times 5 = 20$ is the same as $20 = 4 \times 5$.

In addition, focusing on different ways to express a relationship between unspecified quantities helps develop a basis for making sense of—and solving—equations that otherwise have no real-world context. For example, if a volume P is removed from each of the two volumes Y and S, we can express the result as $Y - P = S - P$. Understanding how to represent removing volume P mathematically is conceptually linked to understanding how to solve an equation such as $27 = 14 + x$, where one might first subtract (remove) 14 from both sides. In this sense, reasoning about relationships between quantities can directly support understanding how to interpret and solve equations.

Functions (Big Idea 5) are also connected with equations in important ways, although they are sometimes studied as separate objects. In particular, we can interpret the solution of an equation geometrically as the point or points at which two graphs intersect. For example, one might view the equation $5(x + 3) = 20$, as the point at which function values for two functions, $y = 5(x + 3)$ and $y = 20$, are equal, or where their graphs intersect. Moreover, because a function rule is the equating of two expressions (for example, $y = 5(x + 3)$ is the equating of the expressions y and $5(x + 3)$), the study of functions relies and builds on the work of understanding equivalence, expressions, and equations. In this sense, functional thinking and reasoning about equations are deeply interconnected.

The fundamental properties govern operations on quantities and thus form the basis for how we operate on equations and functions. In particular, transforming expressions—whether they are used with equations or functions—into equivalent forms requires a deep understanding of the fundamental properties and how to compose and decompose quantities. In this sense, using arithmetic as a context for algebraic thinking is foundational to developing more formal algebraic skills associated with solving equations and interpreting functions.

Big Idea 4

Quantitative reasoning extends relationships between and among quantities to describe and generalize relationships among these quantities.

Big Idea 5

Functional thinking includes generalizing relationships between covarying quantities, expressing those relationships in words, symbols, tables, or graphs, and reasoning with these various representations to analyze function behavior.

The variety of contexts in which algebra occurs suggests that algebra should not be viewed as a separate strand or content area, but rather as a way of thinking that can—and should—be embedded in multiple contexts. Understanding the big ideas discussed in this chapter, as well as how these big ideas are connected, is an essential first step in recognizing algebra across mathematics.

Chapter 2

Connections: Looking Back and Ahead in Learning

A longitudinal approach to algebra intentionally connects core algebraic ideas across prekindergarten through grade 12, giving students the time and experiences that they need to develop a deeper understanding of these ideas. Although the focus of this book is grades 3–5, early algebra should begin at the start of elementary school instruction, if not before. As Kilpatrick, Swafford, and Findell (2001) note, "From the earliest grades of elementary school, students can be acquiring the rudiments of algebra, particularly its representational aspects and the notion of variable and function" (p. 419). In this chapter, we examine how algebraic ideas are connected across the grades, focusing particularly on how ideas in the prekindergarten–grade 2 years lead to and support ideas in grades 3–5, and how these ideas in turn underpin more sophisticated concepts in grades 6–8.

Developing the Symbolic Language of Algebra

Algebra is rooted in a mathematical language that combines operations, variables, and numbers to express mathematical structure and relationships in succinct forms. For example, the commutative property of multiplication is often expressed as $a \times b = b \times a$, for real numbers a and b; or the function $A = \pi r^2$, for non-negative real numbers A and r, might be used to represent the area, A, of a circle with radius r. Moreover, the fundamental properties allow us to operate on these symbolic forms in ways that support mathematical reasoning—as suggested by the discussion of these properties in relation to Big Idea 1. For example, not only can we express an arbitrary odd number as $2n + 1$, where n is any integer, but we can also use the fundamental properties to transform mathematical

Big Idea 1

Addition, subtraction, multiplication and division operate under the same properties in algebra as they do in arithmetic.

expressions to help us reason about what happens when we add two arbitrary odd numbers: Because $(2n + 1) + (2m +1) = (2n + 2m) + (1 + 1) = 2(n + m) + 2 = 2(n + m + 1)$ for integers m and n, we conclude that the sum of any two odd numbers must be even.

Understanding symbolic algebraic notation—particularly, the use of variables—is central to learning algebra. The discussion of Big Idea 3 underscores this deep linguistic aspect of algebra by identifying the different roles that variables might play in a problem. The development of an understanding of variable can begin as early as first grade, or even kindergarten, in the rich and varied experiences that children can have in working with numbers and operations, in developing a relational understanding of equality, in writing equations to represent problem situations, or in exploring functional relationships. Accurately interpreting variables and, ultimately, producing accurate expressions with variables, is a process that develops over time. In this sense, young children benefit from sustained interactions with algebraic notation that begin earlier rather than later. In what follows, we describe some of these connections.

The study of addition and subtraction in prekindergarten through grade 2 can be extended so that fundamental properties or other arithmetic generalizations can be lifted out, identified, and expressed by using natural language and variables. For example, at an early age, children might describe the commutative property of addition in a statement such as, "You can add two numbers in any order," and this can be a springboard for introducing variables in grades 3–5 (or earlier) to express the generalized pattern $a + b = b + a$.

Arithmetic word problems, which often contain unknown amounts to be determined, can also be used to build an understanding of a variable as a fixed but unknown quantity. Consider the following problem:

The Candy Problem

John has 5 pieces of candy. Mary has 3 more pieces than John. How many pieces of candy does Mary have?

Students might initially use manipulatives, drawings, or even their fingers to count on from 5 until they get 8. Once students are able to understand and solve this problem, they can also be guided to represent the mathematics in the problem as $5 + 3 =$ ___ , or subsequently as $5 + 3 = a$, where a represents the number of pieces of candy that Mary has. Moreover, representing simple problems such as these with variables can be a good place to clarify students' understanding that a variable represents a measure or amount, not the object itself (Essential Understanding 3b). Here, as we noted, a represents the *number of pieces* of candy that Mary has, not the candy itself.

As children write equations to represent problem situations, it is easier for them to do this for problems using whole numbers rather

Big Idea 3

Variables are versatile tools that are used to describe mathematical ideas in succinct ways.

For a detailed discussion of the commutative property of addition and its role in flexible computation in the early grades, see *Developing Essential Understanding of Addition and Subtraction for Teaching Mathematics in Prekindergarten–Grade 2* (Caldwell, Karp, and Bay-Williams 2011).

Essential Understanding 3b
A variable represents the measure or amount of an object, not the object itself.

than fractions or decimals, and for addition and subtraction rather than multiplication or division. In this sense, the early elementary grades, where the focus is on adding and subtracting whole numbers, is a natural starting point for developing this skill. Moreover, making connections between word problems and equations not only helps students develop an understanding of a variable as a fixed though unknown number but also starts to build their experience with equations containing variable expressions—common objects in algebra after the elementary grades.

We can extend the development of an understanding of variable further in young children's thinking by removing information from an arithmetic task. Consider the following revised Candy problem, where the number of pieces of candy that John has is now unknown:

> John has some pieces of candy. Mary has 3 more pieces than John. How would you describe the number of pieces of candy that Mary has? (Adapted from Carraher, Schliemann, and Schwartz 2008, p. 238.)

In tasks like this, children begin to grapple with constructing variable expressions. In this particular case, they must struggle with the ambiguity of representing John's unknown amount of candy. Given that most children's mathematical experiences are likely to have focused on arithmetic (and the use of specific numbers), their first inclination is often to specify the number of pieces of candy that John has. It is important to guide their thinking so that they understand that the amount of candy that John has is unknown and is intended to remain unknown and that the solution to the problem is not a specific numerical value.

Indeed, if we expressed the number of pieces of candy that John has as, say, n, then we could express the solution to the problem (that is, the number of pieces of candy that Mary has) as $n + 3$. A unique aspect of such a task is that children are not being asked to solve or compute, but simply express. This represents a critical understanding that children can begin to develop in the context of arithmetic prior to grade 3. Moreover, this understanding can support their development of algebraic ideas through grades 3–5 and into the middle grades and high school, where constructing, operating on, and reasoning with variable expressions are essential parts of algebra.

The notion of a variable as a varying quantity—one of the many ways to interpret a variable, as noted in Essential Understanding 3*a*—can also be introduced prior to grades 3–5 through an exploration of functions. As children explore simple scenarios in which two quantities vary simultaneously—for example, the number of feet in the classroom in relation to the number of students—they can learn

Essential ⬅
Understanding 3*a*
The meaning of
variable *can be*
interpreted in
many ways.

to use variables to express these varying quantities and notice that a variable might take on a range of values. Consider the Handshake problem and figure 2.1, which depicts a function table constructed by one first-grade student to organize her data on the problem:

The Handshake Problem

Suppose that everyone in your group shakes hands with each other once. How many handshakes will there be if there are 2 people in your group? Three people? Four people? Five people? Six people? Organize your information in a function table. Do you see any patterns?

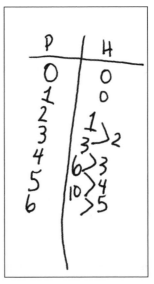

Fig. 2.1. A first grader's function table for the Handshake problem. (Reprinted from Blanton [2008, p. 41].)

In this table, the student expressed the number of people as *P* and the number of handshakes as *H*. Although her understanding of a variable as a varying quantity, as well as her ability to use variables to describe such quantities, is no doubt in its early stages, experiences like this are essential for young children. It is important, for example, that children themselves generate varying data sets and characterize these data sets through notation that uses variables. It is through sustained interactions with such tasks that children begin to construct an understanding of a variable as a varying quantity. Although the choice of letters is arbitrary, teachers should take care that young students do not confuse the object with the quantity—in this case, that they do not assume that *P* represents people rather than the number of people or that *H* represents handshakes rather than the number of handshakes (see Essential Understanding 3*b*).

In grades 3–5, children's understanding and use of variable notation extends from representing varying quantities with variables

➡ **Essential Understanding 3*b***
A variable represents the measure or amount of an object, not the object itself.

and expressing these ideas in organized ways through representations such as function tables, to describing functional relationships in symbolic form. One teacher described how her students represented the relationship they found in the Squares and Vertices problem, discussed in chapter 1 (see fig. 1.11):

> I directed my class to try and form a natural language statement about the function rule they had just discovered. Brandon raised his hand and stated, "You have to take the number of squares and multiply it by three and add one. This will give you the number of vertices." "O.K.," I said. "Let's see if we can revise this sentence to see if we can clean it up a bit." As a class, my students worked in groups to arrive at the conjecture, "Multiply the number of squares by 3 and then add 1." I asked, "Do you think we can make [an equation] from this conjecture?" My students actively went to work. I gave them a few minutes and asked if anyone wanted to come up to the board. Sienna went up to the board and wrote, S $\times$ 3 + 1 = V, where S represented the number of squares and V the number of vertices." (Blanton 2008, p. 44)

Understanding the different roles of variables and how to use variables to describe mathematical ideas succinctly can begin in the early elementary grades and be refined in upper elementary grades. By the middle grades, students can have a richer understanding of variable that is rooted in their experiences in generalizing arithmetic, reasoning about functional relationships, or reasoning quantitatively. In grades 6–8, students need symbolic language skills to help them express mathematical generalizations symbolically, to transform algebraic expressions into equivalent forms, to solve equations, to reason with variable quantities, and to describe and interpret symbolic rules for functional relationships.

Through experiences in the elementary grades with variables in different roles—to express a generalized pattern, represent a fixed but unknown quantity, or represent a varying quantity—students are better prepared to explore more complex mathematical situations with variables. In their more formal study of algebra in the middle grades and beyond, these experiences include encountering a variable as parameter.

In the middle grades, students need to be able to interpret the different roles of variables in linear equations represented in standard forms, such as the point-slope form of a line, often described as $y - y_1 = m(x - x_1)$. In this representation, the variables x and y represent varying quantities, but the variables y_1, x_1, and m represent fixed amounts. Standard forms such as this, which require students to interpret different uses of variables in a single problem situation,

For more information on how teachers might use contexts such as work with blocks to help children in prekindergarten–grade 2 develop an understanding of the fundamental properties, see *Developing Essential Understanding of Number and Numeration for Teaching Mathematics in Prekindergarten–Grade 2* (Dougherty et al. 2010).

The properties of addition and place value are the mathematical foundations for understanding computational procedures for addition and subtraction of whole number—a big idea in *Developing Essential Understanding of Addition and Subtraction for Teaching Mathematics in Prekindergarten–Grade 2* (Caldwell, Karp, and Bay-Williams 2011).

➡️ **Essential Understanding 1e** Generalizations in arithmetic can be derived from the fundamental properties.

are fundamental to the study of algebra in the middle grades and are a precursor to more complex nonlinear functions and their generalized forms encountered in high school algebra.

Generalizing Arithmetic

In the discussion of Essential Understanding 1a, we described how arithmetic serves as a context for algebraic thinking in grades 3–5 by focusing on the fundamental properties. In particular, as children use these properties in computational work, they can learn to notice them, to describe them by using natural language and variables, and to explore why they might be true for the domain of numbers appropriate to their experience.

But children can start to attend to these principles prior to third grade. For example, simple actions such as adding 3 blocks to 4 blocks and 4 blocks to 3 blocks creates a context in which teachers can help children notice that the results are the same regardless of the order of the addends. Children might also notice, for example, a unique characteristic of zero—that is, when zero is added to any number, the result is that number. This can set the stage for a more formal treatment of the additive identity property. Moreover, depending on the extent of these experiences, children who have not yet reached third grade can begin to describe properties of addition (and other arithmetic generalizations) by using natural language, and even variables (Carpenter, Franke, and Levi 2003).

Exploring properties of addition and subtraction prior to third grade sets the stage for similar explorations with multiplication and division in the upper elementary grades. Furthermore, although work with the fundamental properties before third grade might focus on operations on whole numbers, children in upper elementary grades can examine these properties on more complex number domains, such as integers or rational numbers. Ultimately, generalizing fundamental properties in the elementary grades lays the foundation for a formal study of algebra in the middle grades and beyond, where such properties are typically stated in their generalized forms as axioms of real numbers. Without algebraic experiences in the elementary grades, in which children explore the fundamental properties in the meaningful context of operations on numbers and begin to describe them in words or variables, students in the middle grades and high school are likely to be challenged by the abrupt introduction of such properties in their more generalized forms.

We also saw in the discussion of Essential Understanding 1e that other arithmetic generalizations besides the fundamental properties offer fruitful entry points into algebra. These might be generalizations about a string of operations (for example, $b + a - a = b$), or generalizations about classes of numbers such as evens and odds (for example, "The product of an even number and an odd number is even"). Prior to third grade, children can begin to identify and describe number patterns that can support generalizing arithmetic in grades 3–5. For instance, through experiences in combining even or odd numbers of objects, they might observe that an even number of objects can be paired with no objects left over or that an odd number of objects can be paired with one object left over. These observations are not only a critical precursor to expressing an arbitrary even number or odd number, but they also represent aspects of arguments that children might construct to support generalizations about sums or products of even numbers and odd numbers in grades 3–5 and beyond.

Another mathematical practice that supports the learning of algebra involves decomposing quantities in insightful ways to facilitate computation. It is important before third grade that children start to think flexibly about how numbers are composed. Experiences in expressing numbers in multiple ways not only helps children in prekindergarten–grade 2 to think about more efficient computational strategies. Algebraically speaking, it also helps to develop their understanding of equivalence and the fundamental properties.

For example, a second grader might add 35 + 47 by thinking of 35 as 30 + 5 and 47 as 40 + 7, add 30 + 40 to get 70 and 5 + 7 to get 12, and then add 70 + 12 to get 82. Hearing this strategy, a teacher could write "35 + 47 = 30 + 40 + 5 + 7," and ask questions such as, "Does this show how you solved this problem?" or, "Is what I wrote a true equation?" Posing such questions could provide a way to provoke conversations about the fundamental properties used to compute the sum or help children confront misconceptions about the equals sign.

Practices of decomposition can be reinforced in grades 3–5 as children explore the more complex operations of multiplication and division. Taken together, these experiences in decomposing quantities help build a foundation for simplifying algebraic expressions in the middle grades and high school because they can be used to focus attention on the structure of quantities and how they are composed (and hence, decomposed) and how the fundamental properties are used to compose and decompose quantities.

Building Relational Thinking and Quantitative Reasoning

Relational thinking and quantitative reasoning are critical to algebra. They are particularly important in the middle grades and high school, where mathematics explicitly addresses processes such as simplifying expressions and solving equations and inequalities.

These transformational processes of algebra require students to see expressions as objects rather than as a series of computations. As the discussion of Big Idea 2 suggests, this understanding is critical. It enables students who are expressing relationships between these objects in equations or inequalities to reason about them in their entirety rather than focus on only one expression—or one side of an equation or inequality—at a time. The ability to see expressions as objects enables students to reason about what algebraic processes they can use to transform an equation or inequality into a sequence of simpler, equivalent statements from which they can obtain a solution.

The foundations for thinking about expressions as objects and, more broadly, a relational understanding of the equals sign, can begin to grow in the early elementary grades. Prior to third grade, experiences with equations such as $17 + 8 = \_\_\_ + 9$ can help children learn to think relationally about the equals sign and avoid developing the misconception that "=" means "Perform the computation on the left side of the symbol" or "The answer comes next." In particular, children can learn to reason about the equation in its entirety, noting that since 9 is one more than 8, then the unknown amount must be one less than 17—that is, 16. This kind of compensation strategy requires children to focus on structural relationships in both expressions simultaneously, rather than performing specific numerical computations indicated in a particular expression. Without a relational understanding of the equals sign, children are most likely to think that the unknown amount is 25 (that is, $17 + 8$) or 34 (that is, $17 + 8 + 9$). In grades 3–5, students can deepen their relational thinking by applying this reasoning to more complex tasks using multiplication and division. For example, in an equation such as $590 \times 45 = 59 \times \_\_\_$, children would need to see that since 590 is divided by 10 to get 59, 45 must be multiplied by 10 to get the unknown amount.

As we described in our discussion of Essential Understanding 2*b*, a relational understanding of the equals sign is essential for solving algebraic equations. Students who interpret the equals sign to mean "The answer comes next" would be hard-pressed to solve an equation like $3x + 5 = 2x + 15$ and, at best, would probably rely on memorizing procedures to solve such equations. In particular, if they think that $2x$ represents the sum of $3x$ and 5, it would be difficult to understand the relationship expressed by this equation or how one might simplify it. Students in grades 3–5 who have a relational understanding of the equals sign can use the subtraction property of equality to solve equations like $999 + 58 = 958 + b$. These students might say, "I am going to start by subtracting 900 from each side of the equation. Now I just have $99 + 58 = 58 + b$, so b has to be 99." These students are prepared to understand why

Big Idea 2

A mathematical statement that uses an equals sign to show that two quantities are equivalent is called an equation.

Essential Understanding 2*b*
Equations can be reasoned about in their entirety rather than as a series of computations to execute.

subtracting $2x$ from each side of the equation $3x + 5 = 2x + 15$ results in an equivalent, simplified equation.

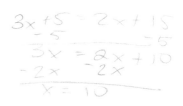

The development of a relational understanding of the equals sign in elementary grades also supports students' understanding of more formal operations on expressions and equations in the middle grades and beyond. Without this understanding, students often attempt to "solve" algebraic expressions rather than transform them into equivalent expressions. For example, when asked to factor $3x + 18$, students without a relational understanding of the equals sign might respond:

$$3x + 18 = 3(x + 6) = 0, \text{ so } x + 6 = 0, \text{ or } x = -6.$$

Instead, relational thinking allows students to transform expressions into equivalent forms and express this equivalence in the form of an equation. In the above example, students would understand that $3x + 18$ is equivalent to $3(x + 6)$, which means that $3x + 18 = 3(x + 6)$ and that no further action (i.e., solving for x) is appropriate. This is also an important prerequisite to solving equations, since solving equations is essentially a sequence of transformations of algebraic expressions determined by reasoning about an equation in its entirety. For example, in the middle grades, students might be asked to solve an equation such as $3(x + 6) = 5x - 3x$. One way of solving this equation symbolically involves re-expressing $3(x + 6)$ as, equivalently, $3x + 18$ and $5x - 3x$ as $2x$ and to understand that these expressions are equivalent. Thus, $3(x + 6) = 5x - 3x$ is equivalent to $3x + 18 = 2x$, and so forth.

In grades 3–5, students can learn to describe simple mathematical expressions algebraically by using variables, and they can model mathematical situations with algebraic equations. As they do this, they can begin to distinguish between mathematical expressions and equations, appreciating that expressions are not intended to be solved and that equations are statements indicating that two expressions are equivalent. With this understanding, students do not need to rely on memorizing algebraic procedures but can focus in the middle grades on building more complex expressions and transforming these expressions into equivalent forms by using the fundamental properties. What is perhaps more important, they can subsequently use these transformational skills to solve algebraic equations.

As the discussion of Big Idea 4 indicates, young children can also learn to reason quantitatively as a foundation for later, more formal work with equations and inequalities. In prekindergarten–grade 2, quantitative reasoning focuses on understanding counting and gaining number flexibility. In these years, children can reason with quantities to explore the idea that when two quantities are equal or unequal, adding an amount to both quantities or

For a discussion of equivalent expressions and equations and solving equations, see *Developing Essential Understanding of Expressions, Equations, and Functions for Teaching Mathematics in Grades 6–8* (Lloyd, Herbel-Eisenmann, and Star forthcoming).

Big Idea 4

Quantitative reasoning extends relationships between and among quantities to describe and generalize relationships among these quantities.

subtracting an amount from both quantities preserves that relationship. Moreover, when *adding* is viewed as joining quantities and *subtracting* is viewed as removing quantities or determining "how much more," then exploring this notion does not require computation but might rely on qualitative comparisons of quantities that are not specified by counts or measurements. For example, young children might compare unspecified quantities, such as the areas of regions or the numbers of uncounted discrete objects, as a way to explore inequality relationships. In the former case, children might compare areas indicated by rectangles of different sizes by superimposing one rectangle on another, drawing a conclusion like, "The area of the orange rectangle is greater than the area of the gray rectangle" (see fig. 2.2). In the latter case, children might qualitatively compare the number of children on a school bus to the number of seats on the bus when they know that the bus has empty seats.

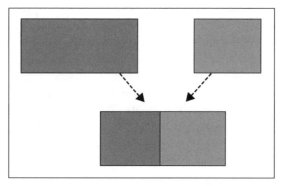

Fig. 2.2. Comparing areas by superimposing one rectangle on another

Reasoning about unspecified quantities helps young children focus on understanding how operations on equivalent or nonequivalent quantities affect the relationship between the quantities and how these relationships might be expressed in natural language, or even with algebraic notation that uses variables. In comparing the number of children on a school bus and the number of seats on the bus, they might first express the relationship in a statement such as, "There are more seats than children." But even earlier than grade 3, children can begin to express and compare equivalent and nonequivalent quantities by using algebraic notation (Dougherty 2008). For example, young children can symbolize this relationship as $A > B$, where A represents the number of seats on the bus and B represents the number of children. Similarly, they might express the relationship between areas of the rectangles as $O > G$, where O represents the area of the orange rectangle and G represents the area of the gray rectangle.

Moreover, as children move into grades 3–5, this type of activity allows them to explore important properties of equality, such as reflexive, symmetric, and transitive properties, and whether or not these properties hold for inequalities. For example, a study of the reflexive property of equality helps children learn to describe and reason about statements such as $A = A$, a precursor to the study of formal algebraic identities in middle and high school. In this sense, the use of quantitative reasoning to develop a deeper understanding of equations and inequalities in elementary grades provides an important foundation for a more formal study of equations and inequalities in transformational algebra in the middle grades and beyond.

Developing Ideas about Functions

The idea of function is deeply threaded through all of mathematics. We are interested in it here because it can serve as an entry point into algebra in the elementary grades. As discussed earlier in this chapter and reflected in Big Idea 5, functional thinking tasks provide opportunities for students to explore the concept of variable as a varying quantity prior to third grade. But they also embody other concepts that are important to algebra and that can be developed in the early elementary grades.

For example, children in the early elementary grades can learn how to generate data from a problem situation and organize these data by using representations such as a function table. Experiences in generating data help children develop increasingly abstract ways to represent quantities and variation in those quantities, and their representations can build from iconic (pictures) to symbolic (variables). In addition, as children attend to *where* data are placed in a function table, they engage in the important work of investigating relationships between quantities, including tracking variation in two quantities simultaneously.

Finally, young children can begin to explore relationships in covarying data, and they might initially use their experiences with arithmetic as a way to make sense of patterns and regularities. For example, using data in their function tables, children might describe recursive patterns in terms of skip counting or correspondence relationships in terms of doubling quantities. Moreover, describing covariational relationships (for example, "As we add one more person to the group, the number of feet in the group increases by two") can be an important conceptual bridge between recursive patterns, which account for variation in only one quantity, and covariational or correspondence relationships, which account for covariation in two quantities.

As children move through grades 3–5, they can build on this foundation by focusing on generalizing correspondence relationships and describing these relationships by using words or variables.

Big Idea 5

Functional thinking includes generalizing relationships between covarying quantities, expressing those relationships in words, symbols, tables, or graphs, and reasoning with these various representations to analyze function behavior.

As Kilpatrick, Swafford, and Findell (2001) describe, children in elementary grades "can observe that over time and across different circumstances, numerical quantities can vary in principled ways.... They can learn about functions by studying how a change in one variable is reflected in the behavior of another" (p. 280). Moreover, the complexity of the functions can increase over time from simple linear functions in the early elementary grades to quadratic and exponential functions in the upper elementary grades. Children in grades 3–5 can also begin to graph functions on a coordinate plane and, through tasks such as the Best Deal problem (see fig. 1.17), can learn to interpret graphs to reason with generalizations in function data and understand problem situations.

Finally, children in grades 3–5 can build on their understanding of recursive patterns and covariation to begin to think informally about how different types of functions change. For example, they might use representations such as tables to notice that some functions grow (or decay) at a constant rate and some do not.

All of this understanding serves as important prerequisite knowledge for the study of algebra in the middle grades, where a more formal treatment of linear functions comes into focus. For example, developing an understanding of different representational tools such as graphs, tables, and variables in elementary grades helps children develop essential skills that can support a deeper analysis of connections among these representations in the middle grades. It can also support the more formal study that occurs in the middle grades of proportional relationships as a special case of linear functions.

For an extended discussion of proportional reasoning and its connections to content related to slope, linear functions, and equations, see *Developing Essential Understanding of Ratios, Proportions, and Proportional Reasoning for Teaching Mathematics in Grades 6–8* (Lobato and Ellis 2010).

Moreover, in the elementary grades, important precursors to a later understanding of the critical notion of slope include developing an understanding of covariation in functions—namely, that change in one variable is coordinated with change in another—and building an informal understanding that different relationships have different patterns of change. In the middle grades, the concept of slope is central to the study of linear functions. It is developed symbolically through the use of variable expressions, graphically through the study of linear functions on the coordinate plane, and numerically through tables of function values representing linear relationships. In the middle grades, students can extend informal investigations in the elementary grades regarding differences in how functions grow or decay to a more focused study of how linear functions grow or decay, based on symbolic, tabular, and graphical representations of their slope. In later years, the study of slope formalizes into the study of rate of change, one of the fundamental concepts of calculus.

Conclusion

Algebraic thinking can begin early in children's mathematical work. It can develop as they explore fundamental properties and other arithmetic relationships; develop an understanding of how quantities relate to each other; learn to work with unknown quantities, equations, and inequalities; and begin to develop a covariational understanding of function. Elementary teachers whose knowledge of mathematical concepts and practices includes the big ideas and essential understandings presented in chapter 1 can capitalize on algebraic opportunities across the mathematics curriculum. These big ideas and essential understandings, along with the connections that we have articulated here in chapter 2, underpin our discussion in chapter 3 of teaching practices and the choice and use of tasks and assessments to support algebraic thinking.

Challenges: Learning, Teaching, and Assessing

The big ideas and essential understandings outlined in chapter 1 describe the mathematical concepts that teachers need to engage students in grades 3–5 in algebraic thinking. Although knowledge of mathematical concepts is important, teachers also need to understand how students learn and how to teach and assess students' understanding most effectively. In this chapter, we provide a brief overview of learning, teaching, and assessing algebraic thinking in grades 3–5.

Practices That Support Algebraic Thinking

Early algebra research shows that children from diverse socioeconomic and educational backgrounds *can* think algebraically. They can develop a relational view of equality; use appropriate representational tools to explore patterns and relationships; identify, symbolize, and reason with functional relationships and properties of numbers and operations; build mathematical arguments that progress from case-based reasoning to more generalized forms that use sophisticated representations; and use unspecified quantities (for example, length or volume) to describe and reason with algebraic relationships.

The impressive forms of algebraic thinking that children exhibit raise a critical question: What classroom practices can teachers use to support it? In what follows, we frame a response to this question first in terms of the Process Standards articulated in *Principles and Standards for School Mathematics* (NCTM 2000), which address problem solving, reasoning and proof, communication, connections, and representation. We then discuss particular types of tasks and assessments that teachers can use to help their students develop an understanding of key ideas of algebra.

Focusing on mathematical processes

The five processes that NCTM emphasizes for school mathematics all play indispensable and interconnected roles in the development of early algebraic thinking. Examination of each process reveals its special contribution to students' understanding of foundational algebraic ideas.

Problem solving

Essential Understanding 1e
Generalizations in arithmetic can be derived from the fundamental properties.

Big Idea 4

Quantitative reasoning extends relationships between and among quantities to describe and generalize relationships among these quantities.

Big Idea 5

Functional thinking includes generalizing relationships between covarying quantities, expressing those relationships in words, symbols, tables, or graphs, and reasoning with these various representations to analyze function behavior.

Early algebra typically involves the use of tasks posed as rich problem-based scenarios that require modeling and reasoning about structure and relationships in mathematical situations. As Carraher and Schliemann (2007) describe, "Problem contexts constitute essential ways to situate and deepen the learning of mathematics and generalizations about quantities and numbers" (p. 690). Fundamental to early algebra is a classroom environment in which children are encouraged to solve problems by developing mathematical conjectures, building arguments to establish or refute these conjectures, and treating established conjectures (generalizations) as both important pieces of shared classroom knowledge and objects of their own reasoning.

The heart of this problem-solving process is generalizing—that is, lifting out and describing the structure inherent in mathematical relationships. In chapter 1, we described a number of mathematical contexts that encourage various kinds of generalizing, including generalizing properties of or relationships in number and operations (Essential Understanding 1e), describing and reasoning with generalized relationships about quantities (Big Idea 4), and finding generalized rules to describe functions (Big Idea 5).

For example, teachers can guide children to build conjectures about sums or products of even numbers and odd numbers. They can help children think about what happens when zero is added to a number, aiding them in developing the generalization that "any number plus zero is that same number" (or $a + 0 = a$, for any real number a). They might help children describe the mathematical relationship between two unspecified quantities, such as the unknown areas of two regions, as a way to facilitate reasoning about equal or unequal quantities. A classroom practice that focuses children's attention on generalizing in these contexts and, more importantly, that situates children *as the ones who build the generalizations*, creates a problem-solving environment of the type that is essential for thinking algebraically.

Finally, for children to be authentically engaged in problem solving, it is important that they be allowed to integrate their own strategies in thinking through problem situations. In a classroom that promotes problem solving, children engage in *anticipatory thinking* (Empson, Levi, and Carpenter 2011)—that is, they approach a mathematical task by first analyzing it to see what relationships might help them reason about it. Anticipatory thinking differs greatly from algorithmic thinking. With algorithmic thinking, a child typically ponders, "What is the first step that I do to solve a problem like this?" In contrast, with anticipatory thinking, a child typically ponders, "What is this problem about, and what relationships do I already understand that will help me solve it?"

Anticipatory thinking is a hallmark of problem solving and is especially crucial to algebraic thinking. When children call on relationships that they already know to help them solve a task, they engage authentically with many different algebraic ideas. Consider a straightforward task such as $81 - 1.3 = b$. In an algorithmic approach, many children will enact a series of steps, such as lining up a decimal point and regrouping to solve this task. In a classroom that promotes problem solving, however, children will first examine the task to see what relationships might help them reason about it. A child might notice that $81 - 1 = 80$, leaving only .3 to be taken away from 80, giving an answer of 79.7.

Because this solution strategy relies on an understanding of subtraction, a child who uses it increases his or her understanding of subtraction with decimal representations. In contrast, enacting the steps of the standard algorithm to solve this problem typically does not cause students to develop their understanding of subtraction. When children bring to the solving of mathematical tasks their understanding of relationships, including relationships in numbers and operations, they increase their general understanding of these relationships.

Reasoning and proof

The activities that underlie early algebra—generalizing and expressing generality—are themselves reasoning processes (Carraher and Schliemann 2007). Generalizations start as conjectured relationships. Because conjectures represent only what one *thinks* might be true, some work has to be done mathematically before a conjecture is accepted as a generalization. As a result, a classroom practice that promotes children's reasoning and proof is critical to early algebra because it pushes children to build mathematical arguments to justify their conjectures. Through this process, conjectures are transformed into generalizations, which can themselves become objects of reasoning about structure and relationships.

Notice that we use the terms *justification* and *argument*, rather than *proof*, which has different connotations and can be ambiguous in the context of elementary grades. The more general notions of justification and argument are more appropriate for elementary grades because they are more inclusive of what children might need

Additional discussion of ways to bridge from addition and subtraction of whole numbers to operations with decimal quantities can be found in *Developing Essential Understanding of Addition and Subtraction for Teaching Mathematics in Prekindergarten–Grade 2* (Caldwell, Karp, and Bay-Williams 2011).

to think about and how they might reason to give an explanation. For example, as we pointed out in our discussion of Essential Understanding 1*b*, the fundamental properties are not statements requiring proof; as axioms, they are assumed to be true. However, it *is* important that children explore these properties and develop convincing arguments to validate their understanding of their truth.

When children make a conjecture that generalizes a relationship, classroom practice that encourages them to think very naturally about *why* or *when* the conjecture might be true supports early algebraic reasoning. For example, classroom practice that helps children move beyond the mechanics of arithmetic with questions that build on one another draws the children's attention to the deeper—*algebraic*—structure underlying arithmetic operations. Consider the effect of a string of questions such as the following: "We talked about why we know that two whole numbers can be multiplied in any order and the product will stay the same, but can two fractions be multiplied in any order? Why might this be true? For what other types of numbers might this conjecture about multiplying in any order be true? Would it work for negative numbers? Does the idea of switching the order of the numbers hold for different operations, such as addition, subtraction, or division?"

Moreover, research has shown that when classroom practice supports reasoning and proof, children learn to build increasingly sophisticated mathematical arguments that progress from testing specific numerical examples to building more generalized arguments (Carpenter, Franke, and Levi 2003). For example, Schifter (2008) describes fourth graders' reasoning in working with representations that grew out of a simple set of subtraction problems: 145 – 100 and 145 – 98. Noticing a connection between the two computations, students wondered if the solution to the second problem would be 2 more or 2 less than the solution to the first, and why. The classroom teacher wrote about students' reasoning:

> Brian was waving his hand in the air, insisting on explaining his thinking, too. He struggled to find the words. "It goes with the problem before," he declared. "It's like you've got this big thing to take away and then you have a littler thing to take away so you have more. Can I draw a picture?" I nodded, and he came up to the blackboard, thought for a while, and then drew a big blob like this [see fig. 3.1]. "See, this is the apple at first," he explained. "And you take some away and have some left. Then you take away 98 grams instead, so it's over here." It appeared to me that Brian had a very clear mental image that was helping him think his way through the problem, but that he was having a hard time communicating it to us. However, his classmates were watching and listening fairly intently. Suddenly, inspired by his presentation, Rebecca said excitedly,

→ Essential Understanding 1*b*
The fundamental properties are essential to computation.

"Yeah, it's like you have this big hunk of bread and you can take a tiny bite or a bigger bite. If you take away smaller, you end up with bigger." "Do you think this will always be true?" I asked. "I think so," she answered. During the discussion up to this point Max had been quiet. Now, inspired by Rebecca's explanation and Brian's picture, he continued further with the thinking that was unfolding. He raised his hand and said, "Yeah, the less you subtract, the more you end up with. AND . . ." he continued with great emphasis, "in fact the thing you end up with is exactly as much larger as the amount less that you subtracted." (Schifter 2008, pp. 77–78; reprinted by permission)

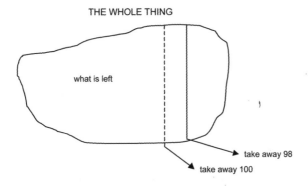

THE WHOLE THING

what is left

take away 98

take away 100

Fig. 3.1. Brian's thoughts on subtraction

The essence of the question here—whether the second computation would be 2 more or 2 less than the first, and why—and the arguments that the students constructed are inherently algebraic. That is, rather than focusing on a computation, students were thinking about a structural relationship between the two problems and reasoning about that relationship in general terms, *independent of specific numbers or operations on those numbers*. Moreover, implicit in this reasoning—and a critical aspect of their algebraic thinking—is the ability to reason about the relationship between two quantities (Big Idea 4). Without this, students would not have been able to simultaneously compare the quantities (known or unknown) represented in the two problems. Instead, they would be limited to an arithmetic approach by computing 145 – 100 and 145 – 98 and comparing differences in the two (numerical) results. Our point is that classroom practice that encourages children to build arguments for their mathematical conjectures supports algebraic understanding.

Teachers can also encourage children to justify functional relationships. Doing so helps children look beyond random patterns in numbers to clues embedded in the context of the problem as a way

Big Idea 4

Quantitative reasoning extends relationships between and among quantities to describe and generalize relationships among these quantities.

➡️ Essential
Understanding 5d
*In working with
functions, several
important types
of patterns or
relationships might
be observed among
quantities that vary
in relation to
each other:
recursive patterns,
covariational
relationships, and
correspondence rules.*

to justify a rule representing the relationships. For instance, in the example of the Squares and Vertices problem, we discussed the process of systematically examining how to find the number of vertices in a configuration of squares. By looking at how to count the number of vertices in configurations consisting of a varying number of squares, we can observe a pattern that justifies the generalized rule, or function (see fig. 1.15 in the discussion of Essential Understanding 5d). Although not all functions lend themselves to this type of reasoning, whenever they do, it is important that teachers' practices support diverse ways to reason about problem situations involving covarying data and the generalizations that arise from them.

Communication

Early algebra thrives in a classroom where discourse about complex ideas is fostered and students are expected to describe their observations about structure and relationships, compare their thinking with that of their peers, defend their conjectured relationships through mathematical arguments, reason publicly about established generalizations, use mathematical language to express their ideas, and so forth.

In such a classroom, teachers focus on two core processes of communication: questioning and listening. Classroom practice based on questioning (rather than telling) can foster the kind of inquiry essential to problem solving and reasoning and proof in algebra. By questioning, the teacher not only tacitly signals to children what kinds of questions are important to think about (for example, "Why do you think this statement is true?" "Is that always going to work?"), but also positions children as responsible for constructing the solutions. In this way, classroom practice creates a context in which children reason about structure in properties of numbers and operations or other arithmetic generalizations, functional relationships, or relationships in generalized quantities.

Not all questions are equal, however. Think about the types of questions that you might ask your students during a typical lesson. Are these questions designed to elicit what students understand and to challenge them to think about important mathematical ideas, or are they designed to assess whether students have retained information that was explained to them and prompt them to enact specific steps in computations? An easy way to decide whether your questions are eliciting students' ideas is to think about the different ways in which your students could answer your questions. If your students might answer a particular question in one of just two ways—either with the right answer or with the wrong answer—then the question is probably not designed to elicit children's ideas. If, by contrast, your students might give one of a variety of valid answers to the question, depending on how the students are thinking, then your question is likely to be useful in assessing your students' ideas.

Questions that support algebraic thinking call on children to explain their thinking, including the strategies that they used to model a problem, the types of representations that they chose, and the salient features of these representations. Such questions can take children beyond simply sharing their ideas to contrasting their thinking with that of others. These questions ask, "How is this strategy similar to or different from other students' strategies?" or, "Which representation helps organize information more succinctly?" They ask children to make their conjectures explicit and to support their conjectures with mathematical arguments: "Do you notice anything that always happens?" "How would you describe your conjecture in words?" "Why do you think your conjecture is always true?" They ask children to develop more mathematical ways of expressing their ideas: "How can you describe your conjecture in symbols (variables)?" "How could you represent this unknown amount without writing out the words?"

Finally, questioning requires listening. A classroom in which the teacher is questioning and actively listening to students' thinking offers an environment in which teachers can make children's algebraic ideas explicit, examine them, and build on them through instruction. Reflect 3.1 explores ideas about productive listening.

Reflect 3.1

Think about your own practice of listening. What do you hear in children's thinking that you might be able to use in ways that could deepen algebraic thinking?

Connections

It is important for teachers to help children make connections among mathematical ideas, understand how their ideas build toward more sophisticated knowledge, and be able to apply their mathematical understanding in contexts outside of mathematics (NCTM 2000). Moreover, when children are expected to solve problems by invoking relationships that they understand, they naturally connect mathematical ideas.

Connections can occur in multiple dimensions. A single early algebra task can give students an opportunity to make important connections *among* the different big ideas presented in chapter 1. For example, in solving the Squares and Vertices problem, children need to write algebraic expressions (for example, $3s + 1$), identify a correspondence relationship, use an equation to represent a problem situation, and interpret the role of a variable as a varying quantity.

Suppose that the task is extended to include the following simple question: "If a configuration of squares has 25 vertices, how many squares would be in the configuration?" In solving a problem

[handwritten margin note: I try to come up w/ a question that goes along w/ what a student has already said to have them explain even more.]

 Big Idea 1

*Addition, subtrac-
tion, multiplication,
and division oper-
ate under the same
properties in algebra
as they do in
arithmetic.*

 Big Idea 2

*A mathematical
statement that uses
an equals sign to
show that two quan-
tities are equivalent
is called an equation.*

 Big Idea 3

*Variables are
versatile tools that
are used to describe
mathematical ideas
in succinct ways.*

Big Idea 5

*Functional thinking
includes general-
izing relationships
between covarying
quantities, express-
ing those relation-
ships in words,
symbols, tables, or
graphs, and reason-
ing with these vari-
ous representations
to analyze function
behavior.*

similar to this, one child, Andrew, reasoned that for the equation $3s + 1 = 25$, he would subtract 1 but that he would need to do this to both quantities for the equation to remain balanced. Applying this understanding, he rewrote the original equation as $3s = 24$. He then took 24 tiles and placed them in groups of 3 and reasoned that since he had 8 groups, the answer had to be 8. For this solution, Andrew needed to use a function to relate the number of vertices to the number of squares, invoke a relational understanding of equality, use the fundamental properties implicitly to simplify the equation, and interpret the variable as a fixed but unknown quan-tity—notions that draw on Big Ideas 1, 2, 3, and 5 regarding funda-mental properties, equations, variables, and functions, respectively. Over time, teachers can help children see these connections among the big ideas by lifting out important associated essential under-standings and guiding them in seeing how these are used within the same task.

Furthermore, if we view algebraic thinking as a habit of mind rather than an isolated set of topics, it is also natural to see it as connected across mathematical domains (for example, arithmetic, geometry, measurement) and even across subject areas other than mathematics in ways that connect important mathematical—algebraic—ideas within a rich web of knowledge. For example, one third-grade teacher designed the Telephone problem to include with a read aloud during her social studies unit on immigration (Soares, Blanton, and Kaput 2006, p. 230):

The Telephone Problem
The second graders at the Jefferson School have raised money to visit the Statue of Liberty. Thirteen friends are planning to go. They are very excited about the trip and worried that they might forget something! On the night before the trip, they call each other to double-check on what they need to bring. Each friend talks to every other friend once. How many phone calls will be made?

As part of solving this task, the students built telephones dur-ing their science unit on sound and used the telephones to collect data about the number of phone calls made for different numbers of friends. The task itself was modeled after the Handshake problem (see p. 70 for a first-grade version of this problem), which had been used previously with this class. As this interdisciplinary example illustrates, teachers can incorporate early algebra tasks in settings outside a traditional mathematics classroom as a way to increase students' access to these ideas. Moreover, the experience of solving a task such as the Telephone problem in a setting other than math-ematics—in this case, a social studies unit—might have unexpected benefits for students who struggle with mathematics.

Representation

Principles and Standards (NCTM 2000) states, "A major responsibility of teachers is to create a learning environment in which students' use of multiple representations is encouraged" (p. 139). In chapter 1, we listed the expectations articulated in the Representation Standard (see p. 55). Although classroom practice that supports the development of representational fluency is important in all of school mathematics, it is especially important in algebra, where diverse representations such as words, variables, tables, pictures, and graphs are the currency for expressing and reasoning with generalizations.

Classroom practice can support children's algebraic thinking through a focus on these diverse representations. For example, children will naturally describe the generalizations that they make by using everyday language. They might describe the multiplicative identity property in a statement such as, "Anytime you multiply a number by 1, you get that same number back." They might express the relationship between the number of children and the number of hands in their classroom by saying, "The number of hands is double the number of children." Or they might characterize a relationship between the areas of two regions with the observation, "The area of this region is greater than the area of that region." In all of these cases, children's representations of their mathematical ideas in words can become the subject of a conversation in which the teacher's purpose is to help the children move toward a symbolic representation, in which they use variables to express their ideas.

In functional thinking tasks, for example, teachers should not only encourage the use of different representations such as words, rules, tables, and graphs to represent functions, but also nurture an understanding of how to navigate among these representations to reason about relationships in covarying data (Big Idea 5). Brizuela and Earnest (2008) describe young children's interpretation of the Best Deal problem (see fig. 1.17) in four different dimensions: (1) written or spoken natural language and instantiations with manipulatives; (2) initial written or algebraic notation, using words or pictures; (3) tables; and (4) graphs. They found that working with different representations in relation to the same problem was critical because it allowed children to use one representation to resolve ambiguities in another. Moreover, they found that moving from one representation to another generated more flexible thinking in children's interpretation of the function. The ability to use different representations to interpret and predict functional behavior—a skill that should begin in the elementary grades—becomes an increasingly important part of algebraic thinking in the middle grades and high school.

Big Idea 5

Functional thinking includes generalizing relationships between covarying quantities, expressing those relationships in words, symbols, tables, or graphs, and reasoning with these various representations to analyze function behavior.

Selecting appropriate tasks and assessments

A discussion of classroom practices that support algebraic thinking requires that we also consider the types of tasks that might be used. An abundance of tasks have been used in early algebra research and can be found in curricular materials. However, rather than list such tasks, which can be readily obtained from publishers, we instead describe more general parameters that should frame our thinking about tasks.

First, early algebra is relatively new as a mathematical focus in the elementary grades. As a result, commonly available curriculum materials and resources might not yet include a focus on developing a strong foundation for algebraic thinking. We return, then, to an earlier point—that algebraic thinking is best viewed as a way of thinking rather than a particular topic area within mathematics. In this sense, the issue is not necessarily what algebra resources are currently available, but where opportunities for algebraic thinking lie in classroom practice. Reflect 3.2 makes this question more immediate and personal.

> ### Reflect 3.2
>
> Think about your own daily classroom practice. Where do you see opportunities for algebraic thinking?

The view that early algebra is a way of thinking about many topics rather than a directed study of particular topics can be transformative because it eliminates dependency on a set of resource materials. Instead, it focuses teachers' attention on identifying instructional opportunities for algebra—whether planned or spontaneous—which can be integrated naturally into instruction. In what follows, we discuss different ways to create opportunities for algebra through task design.

Although early algebra often involves the use of rich open-ended problem situations, simple arithmetic tasks can also be strategically designed to foster algebraic thinking. In chapter 1, we saw how computational problems could leverage thinking about the fundamental properties (Essential Understanding 1b). For example, it is easier to use (and thus, explicitly identify) the distributive property for 4×23 than 4×27, since students typically know 4×3 before they know 4×7. They can use this knowledge to express 4×23 as $4 \times (20 + 3)$, which they can expand by the distributive property to $4 \times 23 = (4 \times 20) + (4 \times 3)$. Teachers can then use this work to focus students' thinking on the property itself.

In other words, teachers who are purposeful in choosing numbers for arithmetic problems can also nurture students' algebraic

➡ **Essential Understanding 1b**
The fundamental properties are essential to computation.

thinking. Schifter's (2008) example of Brian's ideas about subtraction (see pp. 84–85) involved two simple arithmetic tasks: compute 145 – 100 and 145 – 98. But the teacher's strategic choice of numbers allowed children to think algebraically by looking for a structural relationship between the two tasks.

Consider another example. Suppose that you are teaching two-digit multiplication. Computational problems intended to give children practice with the procedure can be designed to achieve a broader—algebraic—purpose. Compare the simple sets of computations in figure 3.2.

Set 1	12 $\times 34$	15 $\times 31$	22 $\times 18$	10 $\times 25$
Set 2	12 $\times 34$	34 $\times 12$	18 $\times 16$	16 $\times 18$

Fig. 3.2. Using computational tasks to support algebraic thinking

Set 1 contains a series of computations that would allow children to practice two-digit multiplication. Although this is also true of the tasks in set 2, the choice of factors in these tasks is designed to help the teacher draw children's thinking to how multiplication behaves when the same two numbers are multiplied in a different order. In this sense, the arithmetic tasks in set 2 offer an important entry point into a conversation by means of which children can develop an understanding of the commutative property of multiplication, express this property in words or variables, think about the domain of numbers for which this generalization might be true, and develop an argument explaining why it might be true for a broad class of numbers (for example, all whole numbers).

Another approach to bringing algebra into arithmetic problems is through what we might call "the algebraic use of number." This approach includes choosing numbers for an arithmetic task that are sufficiently large or complex that students are not able to use arithmetic to solve the problem but instead must think about its underlying structure. For example, one third-grade teacher described how she transformed an arithmetic task into an algebraic task by asking students to determine whether the sum of two large even numbers was even or odd. By using three-digit whole numbers, which the students had no arithmetic procedure for adding, the teacher forced them to think about the structural properties of even numbers as they reasoned about the sum (Blanton 2008).

Consider another example. Think back to the problem in chapter 1 involving JaeQwan's flowerpots, each of which takes $3/4$ of

More examples of integrating algebraic thinking into computation and suggestions for which properties are appropriate at specific grade levels can be found in the Common Core State Standards for Mathematics (Common Core State Standards Initiative 2010).

a pound of clay to make. Recall our extension of the problem, in which we asked how many pots JaeQwan could make if she had 4,327 pounds of clay. Using a large amount of clay helps children appreciate that their arithmetic strategies are inefficient and that equations can serve as useful tools for modeling and solving a problem situation.

Another useful strategy is to convert known information in an arithmetic task to unknown information. In chapter 2, we presented the Candy problem:

> John has 5 pieces of candy. Mary has 3 more pieces than John. How many pieces of candy does Mary have?

As we described, by making the number of pieces of candy John has an unknown amount ("John has some pieces of candy"), we create a task whose solution—the number of pieces of candy Mary has—is an algebraic expression. This type of task pushes an important conceptual transition in children's algebraic thinking, helping them to confront the ambiguity of solutions that are not specific numerical values (Carraher, Schliemann, and Schwartz 2008).

In addition, tasks where the measures of quantities such as mass, volume, or area are intentionally left unspecified help children learn to reason algebraically about relationships among these quantities (Dougherty 2010). For example, when young children are asked to compare the volumes of liquids in two beakers without knowing the specific amounts, the conceptual challenge that they face in representing an unknown volume of liquid is similar to that of describing the unknown amount of candy in the earlier example.

We also noted the limitation of arithmetic equations written in a standard format, with all computation on the left of the equals sign and the unknown on the right, as in $24 \times 31 = $ ___. We observed that deviating from this format can help students develop a relational understanding of the equals sign, an achievement that is not trivial. The most important thing that teachers can do to support a relational understanding of the equals sign (Essential Understanding 2a) is to make sure that children regularly see the equals sign in different places in equations, as in equations like $23 = $ ___ $+ 15$, which are not in a standard format, and $12 \times 5 = 4 \times $ ___, with operations on both sides of the equals sign.

Finally, by varying a fixed quantity, arithmetic tasks can often be modified to create functional thinking tasks. For example, figure 3.3 shows an arithmetic version of the Squares and Vertices problem. By simply varying the number of squares, we get the functional thinking task described in chapter 1 (see fig. 1.11).

Our point here is that tasks that support children's algebraic thinking can arise through simple arithmetic problems that have been transformed in subtle but strategic ways to build opportunities

→ **Essential Understanding 2a**
The equals sign is a symbol that represents a relationship of equivalence.

arithmetic

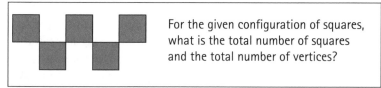

For the given configuration of squares, what is the total number of squares and the total number of vertices?

Fig. 3.3. An arithmetic version of the Squares and Vertices problem

for generalizing and for expressing and reasoning with generalizations. The possibilities include—

- choosing numbers strategically so that arithmetic tasks can prompt an exploration of fundamental properties and other generalizations;

- choosing numbers that are large enough so that children cannot use arithmetic to solve the problem, but must instead focus on the problem's algebraic structure;

- writing equations in nonstandard formats so that children develop a relational understanding of equality;

- making known information in a particular problem unknown so that children think about how to express unknown quantities; and

- varying a previously fixed quantity to create a functional thinking task.

Assessing Algebraic Thinking

Assessment should not be viewed as an endpoint, but as part of a cycle of learning by which children gradually increase their algebraic understanding. Moreover, assessment of children's understanding of algebraic ideas needs to include a variety of informal and formal ways to understand their thinking.

Assessments, like early algebra tasks, should include rich problem-based situations that require higher-order thinking, are often open-ended in nature, and benefit from meaningful classroom discourse rooted in argumentation and justification. Informally, teachers should frequently question students regarding their strategies, ideas, conjectures, and justifications, not only as a part of regular instruction, but also as part of their assessment practice. Formal assessments should reflect the nature of early algebra tasks, like those tasks described here, and not be based on rote recall of knowledge.

And, like tasks used in classroom instruction, assessments should be designed to probe and reveal potential difficulties in students' thinking. For example, a simple task such as $8 + 4 = \_\_\_ + 5$ can reveal important information about children's understanding of

Big Idea 4

Quantitative reasoning extends relationships between and among quantities to describe and generalize relationships among these quantities.

Big Idea 5

Functional thinking includes generalizing relationships between covarying quantities, expressing those relationships in words, symbols, tables, or graphs, and reasoning with these various representations to analyze function behavior.

the equals sign (Carpenter, Franke, and Levi 2003). Finally, assessment—and classroom instruction more broadly—should include the variety of entry points into algebra described in the big ideas of this book. Although any one of these Big Ideas (e.g., Big Idea 4 regarding quantitative reasoning or Big Idea 5 regarding functional thinking) represents an essential component of algebra understanding, it is collective attention to all of these domains of algebra that can provide children with the connected algebra knowledge that they will need for the middle grades and beyond.

Learning to Think Algebraically

Early algebra can help children develop a deep conceptual foundation for generalizing mathematical ideas and for representing, justifying, and reasoning with these generalizations. A fundamental part of this development involves helping children connect informal and formal ways of representing their algebraic thinking. Formal expressions lack depth and grounding if children cannot relate them to informal expressions. Conversely, informal expressions lack power and generalizability if they are not ultimately related to formal expressions. The kinds of classroom practices described earlier, as well as the assessments and tasks that might be used in those practices, are fundamental to building rich algebra experiences in the elementary grades to support this learning. In what follows, we briefly describe ways in which learning early algebra can help students master some of the challenges associated with later algebra.

As children begin to identify fundamental properties and other arithmetic generalizations, describing these in words and symbols and explicitly attending to their use in computations or reasoning with them in mathematical arguments, they develop a conceptual basis for understanding the formal use of these arithmetic generalizations in later grades. For example, explicitly understanding how the distributive property works in transforming a quantity such as $3 \times (40 + 6)$ into $3 \times 40 + 3 \times 6$ can help students understand why $9(x + 5) = 9x + 45$ and lead to a formal understanding that $a(b + c) = ab + ac$, not $ab + c$, or that $3n + 4n = (3 + 4)n = 7n$, while $3n + 4 \neq 7n$.

Without such experiences in arithmetic, students run the risk of encountering the fundamental properties as formal axioms in high school algebra without developing any understanding of their meaning. Similarly, arithmetic generalizations such as those about products of even numbers and odd numbers, or general understandings such as "–x is not necessarily a negative number," can seem too abstract to use as tools for reasoning and building arguments if students do not develop them through their own experiences in arithmetic.

As we have noted earlier, a relational understanding of equality is critical for addressing some of the difficulties that students have with equations. Students who hold a misconception about equality may incorrectly conclude that an equation such as 6 = 6 is not true since it presents no computation to perform, or that 6 = 3 + 3 is not true because the indicated sum is to the right of the equals sign, or that 8 + 4 = 12 + 5 is true since 8 + 4 = 12. This misconception about equality may arise from the fact that equations are more often used to prompt students to compute rather than to represent or reason about mathematical relationships. As we described earlier, one way to develop a relational understanding of equality is for children to see and solve equations written in non-standard forms, with, for example, operations on both sides of the equation or unknown amounts appearing on either side.

Experiences in early algebra support children as they make the transition from the use of natural language into more symbolic ways of expressing mathematical generalizations. In learning algebra, it is important that children first describe their generalizations in their own words, since this experience can be a springboard for introducing variables as shorthand for describing ideas that they have already examined (Schoenfeld and Arcavi 1988).

Additionally, the use of numbers as *quasi-variables* can serve as an important bridge between arithmetic and symbolic algebraic expressions. Fujii (2003) defines a *quasi-variable* as a number sentence or group of number sentences that convey an underlying mathematical relationship that is true regardless of the numbers used. For example, a student may compute the sum 73 + 19 by first subtracting 1 from 73 and adding it to 19 to produce the equivalent, easier, sum 72 + 20. In particular, the student is employing $73 + 19 = (73 - 1) + (19 + 1)$ as a quasi-variable for the algebraic relation $a + 19 = (a - 1) + 20$, or more generally, $a + b = (a - c) + (b + c)$, though without using formal algebraic thought or language. To the extent that the child sees this type of strategy as valid no matter what the addends are, he or she is using numbers as quasi-variables. In this sense, numbers can help link arithmetic thinking to algebraic thinking.

It is important for children to have numerous experiences with variables to understand their different roles as well as what a literal symbol (a letter of the alphabet) represents. For example, as children in grades 3–5 write equations to represent and solve arithmetic word problems, they can develop a notion of a variable as a fixed but unknown number (Essential Understanding 3*a*). Examining functional relationships that describe how quantities co-vary gives children experiences with a variable as a varying quantity. Generalizing fundamental properties, such as the commutative property of multiplication expressed as $a \times b = b \times a$, gives them a context for exploring a variable as a generalized number. Although children might not explicitly name these different uses of variables

➡️ **Essential Understanding 3a**
The meaning of variable *can be interpreted in many ways.*

in elementary grades, their experiences provide a critical foundation for formalizing the different roles of variables in later grades.

Moreover, the elementary grades are important years for children to develop an accurate understanding of what a variable actually represents in a problem situation. As we noted earlier, students often mistakenly suppose that a variable represents an object. For example, if variable *A* represents the area of a triangle, students might interpret *A* to be the triangle itself, rather than its area. Repeated experiences with the use of a variable, where attention is given to its proper meaning, can help students develop an appropriate understanding for what that variable represents.

Furthermore, early algebra not only helps children learn to represent ideas symbolically, it also helps them construct meaning for the relationships that they express. For example, as children solve function tasks by modeling problem situations, identifying and organizing data generated from these situations, describing relationships in data, representing these relationships in multiple ways, and interpreting the relationships to solve problems, they develop an understanding of the meaning behind the symbolic rules—their more common representation in later grades—and how the problem situation, data, and representations of the relationship are connected with one another. Moreover, function tasks that involve geometric patterns can provide children with important visual cues that help them see patterns of change and identify connections between symbolic rules (for example, $v = 3s + 1$) and the problem context. In contrast, students who are given formalized rules as a starting point for functional thinking have not had opportunities to construct meaning for these symbols and how they might arise from a particular context.

Finally, just as children's development of the idea of a variable benefits by starting with their natural language, learning to represent relationships between quantities benefits by first reasoning qualitatively about these relationships. Lochhead and Mestre (1988) describe three levels through which children might progress as they reason about relationships. For example, consider the following problem: "At the class bake sale, the students sold three times as many cookies as cupcakes." At a *qualitative level*, children might first analyze the relationship holistically by considering the question, "Were more cookies or cupcakes sold?" At a *quantitative level*, they might then think about the relationship in terms of specific numbers of items. For example, if 10 cupcakes were sold, how many cookies were sold? If 15 cupcakes were sold, how many cookies were sold? Finally, at a *conceptual level*, they might use these questions to yield data that they could organize in an appropriate representation (for example, function tables) to help them describe a relationship between the number of cookies and the number of cupcakes that were sold at the bake sale.

An implicit aspect of these levels is the *explicit* inclusion of both the number of cupcakes and the number of cookies in all of them. This is important to keep in mind, especially for functional thinking tasks. The type of relationship that children might identify in function data—recursive, covariational, or correspondence (see Essential Understanding 5*d*)—often occurs hierarchically. That is, children will typically first identify a recursive pattern in the function values (values for the dependent variable) because it is easier to focus on variation in a single sequence of data.

However, it is worth noting that research suggests that children might use a hybrid approach that draws on both recursive and covariational ways of thinking. For example, they might look for an additive relationship between the value of the independent variable and the function value, and then use this relationship to find the function value (for an example, see Martinez and Brizuela [2006]). Indeed, it is important to shift their attention to how two quantities vary in relation to each other, and identifying a covariational relationship in function data is a good starting point for drawing their attention away from recursive patterns. For this purpose, keeping both quantities as an explicit part of the problem helps children begin to attend to two quantities, and the variables that represent them, from the outset. This can help children avoid developing an entrenched focus on recursive patterns and can set the stage for learning how to identify correspondence relationships.

Essential ←
Understanding 5*d*
In working with functions, several important types of patterns or relationships might be observed among quantities that vary in relation to each other: recursive patterns, covariational relationships, and correspondence rules.

Conclusion

In this book, we have identified the big ideas and essential understandings of early algebra that elementary teachers need to know. We have used these to think about how the corresponding algebraic thinking that is appropriate for grades 3–5 is connected to algebraic ideas that children develop naturally in earlier elementary grades and extends into the understanding that they need to have in the middle grades. We have closed with a characterization of the kinds of classroom practices, and associated tasks and assessments, that can support children's algebraic understanding, and how instruction that focuses on algebra in grades 3–5 can help move students beyond some of the challenges that they might face if they do not have these early experiences in algebra.

What seems clear is that when elementary teachers provide children with rich, meaningful opportunities to solve mathematical problems that lead them to focus on generalizing and reasoning with generalizations, to justify their mathematical claims with appropriate arguments, to communicate their ideas in different forms, to build connections among these ideas across different mathematical domains and outside of mathematics, and to represent mathematical generalizations in multiple ways and navigate among these representations, they help children develop an essential understanding of algebra that will support their algebraic readiness for later grades. In this context, the ideas described here are intended to provide a conceptual road map for building elementary teachers' understanding of algebra and, ultimately, that of the children whom they teach.

References

Anku, Sitsofe Enyoman. "From Matching to Mapping: Connecting English and Mathematics." *Mathematics Teaching in the Middle School* 2 (February 1997): 270–72.

Barnett-Clarke, Carne, William Fisher, Rick Marks, and Sharon Ross. *Developing Essential Understanding of Rational Numbers for Teaching Mathematics in Grades 3–5.* Essential Understanding Series. Reston, Va.: National Council of Teachers of Mathematics, 2010.

Blanton, Maria. *Algebra and the Elementary Classroom: Transforming Thinking, Transforming Practice.* Portsmouth, N.H.: Heinemann, 2008.

Blanton, Maria, and James Kaput. "Characterizing a Classroom Practice That Promotes Algebraic Reasoning." *Journal for Research in Mathematics Education* 36 (November 2005): 412–46.

Brizuela, Bárbara, and Darell Earnest. "Multiple Notational Systems and Algebraic Understandings: The Case of the 'Best Deal' Problem." In *Algebra in the Early Grades*, edited by James Kaput, David Carraher, and Maria Blanton, pp. 273–301. Mahwah, N.J.: Lawrence Erlbaum Associates, 2008.

Caldwell, Janet H., Karen Karp, and Jennifer M. Bay-Williams. *Developing Essential Understanding of Addition and Subtraction for Teaching Mathematics in Prekindergarten–Grade 2.* Essential Understanding Series. Reston, Va.: National Council of Teachers of Mathematics, 2011.

Carpenter, Thomas P., Megan Loef Franke, and Linda Levi. *Thinking Mathematically: Integrating Arithmetic and Algebra in Elementary School.* Portsmouth, N.H.: Heinemann, 2003.

Carraher, David, and Analucia Schliemann. "Early Algebra." In *Second Handbook of Research on Mathematics Teaching and Learning*, edited by Frank K. Lester, pp. 669–705. Charlotte, N.C.: Information Age; Reston, Va.: National Council of Teachers of Mathematics, 2007.

Carraher, David, Analucia Schliemann, and Judah Schwartz. "Early Algebra Is Not the Same as Algebra Early." In A*lgebra in the Early Grades*, edited by James Kaput, David Carraher, and Maria Blanton, pp. 235–72. Mahwah, N.J.: Lawrence Erlbaum Associates, 2008.

Common Core State Standards Initiative. *Common Core State Standards for Mathematics. Common Core State Standards (College- and Career-Readiness Standards and K–12 Standards in English Language Arts and Math).* Washington, D.C.: National Governors Association Center for Best Practices and

the Council of Chief State School Officers, 2010. http://www
.corestandards.org.

Cooney, Thomas P., Sybilla Beckmann, and Gwendolyn M. Lloyd.
*Developing Essential Understanding of Functions for Teaching
Mathematics in Grades 9–12*. Essential Understanding Series.
Reston, Va.: National Council of Teachers of Mathematics, 2010.

Dougherty, Barbara J. "Measure Up: A Quantitative View of Early
Algebra." In *Algebra in the Early Grades*, edited by James J.
Kaput, David W. Carraher, and Maria L. Blanton, pp. 389–412.
New York: Lawrence Erlbaum Associates, 2008.

———. "A Davydov Approach to Early Mathematics." In Future
Curricular Trends in Algebra and Geometry, edited by Zalman
Usiskin, Kathleen Andersen, and Nicole Zotto, pp. 63–69.
Charlotte, N.C.: Information Age Publishing, 2010.

Dougherty, Barbara J., Alfinio Flores, Everett Louis, and Catherine
Sophian. *Developing Essential Understanding of Number and
Numeration for Teaching Mathematics in Prekindergarten–
Grade 2*. Essential Understanding Series. Reston, Va.: National
Council of Teachers of Mathematics, 2010.

Ellis, Amy B. "Algebra in the Middle School: Developing Functional
Relationships through Quantitative Reasoning." In *Early
Algebraization: A Global Dialogue from Multiple Perspectives*,
edited by Jinfa Cai and Eric Knuth. New York: Springer, in
press.

Empson, Susan B., and Linda L. Levi. *Extending Children's
Mathematics: Fractions and Decimals*. Portsmouth, N.H.:
Heinemann, 2011.

Empson, Susan B., Linda L. Levi, and Thomas P. Carpenter. "The
Algebraic Nature of Fractions: Developing Relational Thinking
in Elementary School." In *Early Algebraization: A Global
Dialogue from Multiple Perspectives*, edited by Jinfa Cai and
Eric Knuth. New York: Springer, 2011.

Fujii, Toshiakira. "Probing Students' Understanding of Variables
through Cognitive Conflict Problems: Is the Concept of
a Variable So Difficult for Students to Understand?" In
*Proceedings of the 27th Conference of the International Group
for the Psychology of Mathematics Education (PME)*, vol. 1,
edited by Neil A. Pateman, Barbara J. Dougherty, and Joseph T.
Zilliox, pp. 49– 65. Honolulu: PME, 2003.

Kaput, James. "What Is Algebra? What Is Algebraic Reasoning?"
In *Algebra in the Early Grades*, edited by James Kaput, David
Carraher, and Maria Blanton, pp. 5–17. Mahwah, N.J.: Lawrence
Erlbaum Associates, 2008.

Kaput, James, David Carraher, and Maria Blanton, eds. *Algebra in
the Early Grades*. Mahwah, N.J.: Lawrence Erlbaum Associates,
2008.

Kieran, Carolyn. "The Changing Face of School Algebra." In *8th International Congress on Mathematical Education, Selected Lectures*, edited by C. Alsina, J. Alvarez, B. Hodgson, C. Laborde, and A. Pérez, pp. 271–286. Seville, Spain: S.A.E.M. Thales, 1996.

Kilpatrick, Jeremy, Jane Swafford, and Bradford Findell, eds. *Adding It Up*: *Helping Children Learn Mathematics*. Washington, D.C.: National Academy Press, 2001.

Lannin, John, Amy B. Ellis, and Rebekah Elliott. *Developing Essential Understanding of Mathematical Reasoning for Teaching Mathematics in Prekindergarten–Grade 8*. Essential Understanding Series. Reston, Va.: National Council of Teachers of Mathematics, forthcoming.

Lobato, Joanne, and Amy B. Ellis. *Developing Essential Understanding of Ratios, Proportions, and Proportional Reasoning for Teaching Mathematics in Grades 6–8*. Essential Understanding Series. Reston, Va.: National Council of Teachers of Mathematics, 2010.

Lloyd, Gwendolyn, Beth Herbel-Eisenmann, and Jon Star. *Developing Essential Understanding of Expressions, Equations, and Functions for Teaching Mathematics in Grades 6–8*. Essential Understanding Series. Reston, Va.: National Council of Teachers of Mathematics, forthcoming.

Lochhead, Jack, and Jose Mestre. "From Words to Algebra: Mending Misconceptions." In *The Ideas of Algebra, K–12*, 1988 Yearbook of the National Council of Teachers of Mathematics (NCTM), edited by Arthur F. Coxford, pp. 127–35. Reston, Va.: NCTM, 1988.

Martinez, Mara, and Bárbara Brizuela. "A Third Grader's Way of Thinking about Linear Function Tables." *Journal of Mathematical Behavior* 25 (January 2006): 285–98.

National Council of Teachers of Mathematics (NCTM). *Principles and Standards for School Mathematics*. Reston, Va.: NCTM, 2000.

———. *Curriculum Focal Points for Prekindergarten through Grade 8 Mathematics: A Quest for Coherence*. Reston, Va.: NCTM, 2006.

———. *Focus in High School Mathematics: Reasoning and Sense Making*. Reston, Va.: NCTM, 2009.

———. *Guiding Principles for Mathematics Curriculum and Assessment*. Reston, Va.: NCTM, 2009. http://www.nctm.org/standards/content.aspx?id=23273.

Otto, Albert Dean, Janet H. Caldwell, Cheryl Ann Lubinski, and Sarah Wallus Hancock. *Developing Essential Understanding of Multiplication and Division for Teaching Mathematics in Grades 3–5*. Essential Understanding Series. Reston, Va.: National Council of Teachers of Mathematics, 2011.

Rivera, Ferdinand, and Joanne Rossi Becker. "Figural and Numerical

Modes of Generalizing in Algebra." *Mathematics Teaching in the Middle School* 11 (November 2005): 198–203.

Schifter, Deborah. "Representation-Based Proof in the Elementary Grades." In *Teaching and Learning Proof across Grades K–16*, edited by Despina A. Stylianou, Maria L. Blanton, and Eric J. Knuth, pp. 71–86. Mahwah, N.J.: Lawrence Erlbaum Associates, 2008.

Schoenfeld, Alan H., and Abraham Arcavi. "On the Meaning of Variable." *Mathematics Teacher* 81 (September 1988): 420–27.

Smith, Erick. "Stasis and Change: Integrating Patterns, Functions, and Algebra throughout the K–12 Curriculum." In *A Research Companion to Principles and Standards of School Mathematics*, edited by Jeremy Kilpatrick, W. Gary Martin, and Deborah Schifter, pp. 136–50. Reston, Va.: National Council of Teachers of Mathematics, 2003.

Smith, Erick, and Jere Confrey. "Multiplicative Structures and the Development of Logarithms: What Was Lost by the Invention of Function." In *The Development of Multiplicative Reasoning in the Learning of Mathematics*, edited by Guershon Harel and Jere Confrey, pp. 333–64. Albany, N.Y.: State University of New York Press, 1994.

Soares, June, Maria Blanton, and James Kaput. "Thinking Algebraically across the Elementary School Curriculum." *Teaching Children Mathematics* 12 (January 2006): 228–35.

Thompson, Pat. "Quantitative Concepts as a Foundation for Algebraic Reasoning: Sufficiency, Necessity, and Cognitive Obstacles." In *Proceedings of the Annual Conference of the International Group for the Psychology of Mathematics Education*, edited by Merlyn J. Behr, Carole B. Lacampagne, and Margariete Montague Wheeler, pp. 163–70. Dekalb, Ill.: Northern Illinois University, 1988.